Probability: A Very Short Introduction

VERY SHORT INTRODUCTIONS are for anyone wanting a stimulating and accessible way into a new subject. They are written by experts, and have been translated into more than 45 different languages.

The series began in 1995, and now covers a wide variety of topics in every discipline. The VSI library now contains over 500 volumes—a Very Short Introduction to everything from Psychology and Philosophy of Science to American History and Relativity—and continues to grow in every subject area.

Titles in the series include the following:

AFRICAN HISTORY John Parker and Richard Rathbone
AGEING Nancy A. Pachana
AGNOSTICISM Robin Le Poidevin
AGRICULTURE Paul Brassley and Richard Soffe
ALEXANDER THE GREAT Hugh Bowden
ALGEBRA Peter M. Higgins
AMERICAN HISTORY Paul S. Boyer
AMERICAN IMMIGRATION David A. Gerber
AMERICAN LEGAL HISTORY G. Edward White
AMERICAN POLITICAL HISTORY Donald Critchlow
AMERICAN POLITICAL PARTIES AND ELECTIONS L. Sandy Maisel
AMERICAN POLITICS Richard M. Valelly
THE AMERICAN PRESIDENCY Charles O. Jones
AMERICAN SLAVERY Heather Andrea Williams
THE AMERICAN WEST Stephen Aron
AMERICAN WOMEN'S HISTORY Susan Ware
ANAESTHESIA Aidan O'Donnell
ANARCHISM Colin Ward
ANCIENT EGYPT Ian Shaw
ANCIENT GREECE Paul Cartledge
THE ANCIENT NEAR EAST Amanda H. Podany
ANCIENT PHILOSOPHY Julia Annas
ANCIENT WARFARE Harry Sidebottom
ANGLICANISM Mark Chapman
THE ANGLO-SAXON AGE John Blair
ANIMAL BEHAVIOUR Tristram D. Wyatt
ANIMAL RIGHTS David DeGrazia
ANXIETY Daniel Freeman and Jason Freeman
ARCHAEOLOGY Paul Bahn
ARISTOTLE Jonathan Barnes
ART HISTORY Dana Arnold
ART THEORY Cynthia Freeland
ASTROPHYSICS James Binney
ATHEISM Julian Baggini
THE ATMOSPHERE Paul I. Palmer
AUGUSTINE Henry Chadwick
THE AZTECS Davíd Carrasco
BABYLONIA Trevor Bryce
BACTERIA Sebastian G. B. Amyes
BANKING John Goddard and John O. S. Wilson
BARTHES Jonathan Culler
BEAUTY Roger Scruton
THE BIBLE John Riches
BLACK HOLES Katherine Blundell
BLOOD Chris Cooper
THE BODY Chris Shilling
THE BOOK OF MORMON Terryl Givens
BORDERS Alexander C. Diener and Joshua Hagen
THE BRAIN Michael O'Shea
THE BRICS Andrew F. Cooper
BRITISH POLITICS Anthony Wright

BUDDHA Michael Carrithers
BUDDHISM Damien Keown
BUDDHIST ETHICS Damien Keown
BYZANTIUM Peter Sarris
CANCER Nicholas James
CAPITALISM James Fulcher
CATHOLICISM Gerald O'Collins
CAUSATION Stephen Mumford and Rani Lill Anjum
THE CELL Terence Allen and Graham Cowling
THE CELTS Barry Cunliffe
CHEMISTRY Peter Atkins
CHILD PSYCHOLOGY Usha Goswami
CHINESE LITERATURE Sabina Knight
CHOICE THEORY Michael Allingham
CHRISTIAN ART Beth Williamson
CHRISTIAN ETHICS D. Stephen Long
CHRISTIANITY Linda Woodhead
CIRCADIAN RHYTHMS Russell Foster and Leon Kreitzman
CITIZENSHIP Richard Bellamy
CLASSICAL MYTHOLOGY Helen Morales
CLASSICS Mary Beard and John Henderson
CLIMATE Mark Maslin
CLIMATE CHANGE Mark Maslin
COGNITIVE NEUROSCIENCE Richard Passingham
THE COLD WAR Robert McMahon
COLONIAL AMERICA Alan Taylor
COLONIAL LATIN AMERICAN LITERATURE Rolena Adorno
COMBINATORICS Robin Wilson
COMMUNISM Leslie Holmes
COMPLEXITY John H. Holland
THE COMPUTER Darrel Ince
COMPUTER SCIENCE Subrata Dasgupta
CONFUCIANISM Daniel K. Gardner
CONSCIOUSNESS Susan Blackmore
CONTEMPORARY ART Julian Stallabrass
CONTEMPORARY FICTION Robert Eaglestone
CONTINENTAL PHILOSOPHY Simon Critchley
CORAL REEFS Charles Sheppard
CORPORATE SOCIAL RESPONSIBILITY Jeremy Moon
COSMOLOGY Peter Coles
CRIMINAL JUSTICE Julian V. Roberts
CRITICAL THEORY Stephen Eric Bronner
THE CRUSADES Christopher Tyerman
CRYSTALLOGRAPHY A. M. Glazer
DADA AND SURREALISM David Hopkins
DANTE Peter Hainsworth and David Robey
DARWIN Jonathan Howard
THE DEAD SEA SCROLLS Timothy Lim
DECOLONIZATION Dane Kennedy
DEMOCRACY Bernard Crick
DEPRESSION Jan Scott and Mary Jane Tacchi
DERRIDA Simon Glendinning
DESIGN John Heskett
DEVELOPMENTAL BIOLOGY Lewis Wolpert
DIASPORA Kevin Kenny
DINOSAURS David Norman
DREAMING J. Allan Hobson
DRUGS Les Iversen
DRUIDS Barry Cunliffe
THE EARTH Martin Redfern
EARTH SYSTEM SCIENCE Tim Lenton
ECONOMICS Partha Dasgupta
EGYPTIAN MYTH Geraldine Pinch
EIGHTEENTH-CENTURY BRITAIN Paul Langford
THE ELEMENTS Philip Ball
EMOTION Dylan Evans
EMPIRE Stephen Howe
ENGLISH LITERATURE Jonathan Bate
THE ENLIGHTENMENT John Robertson
ENVIRONMENTAL ECONOMICS Stephen Smith
ENVIRONMENTAL POLITICS Andrew Dobson
EPICUREANISM Catherine Wilson
EPIDEMIOLOGY Rodolfo Saracci
ETHICS Simon Blackburn
THE ETRUSCANS Christopher Smith
EUGENICS Philippa Levine
THE EUROPEAN UNION John Pinder and Simon Usherwood
EVOLUTION Brian and Deborah Charlesworth

EXISTENTIALISM Thomas Flynn
THE EYE Michael Land
FAMILY LAW Jonathan Herring
FASCISM Kevin Passmore
FEMINISM Margaret Walters
FILM Michael Wood
FILM MUSIC Kathryn Kalinak
THE FIRST WORLD WAR Michael Howard
FOOD John Krebs
FORENSIC PSYCHOLOGY David Canter
FORENSIC SCIENCE Jim Fraser
FORESTS Jaboury Ghazoul
FOSSILS Keith Thomson
FOUCAULT Gary Gutting
FREE SPEECH Nigel Warburton
FREE WILL Thomas Pink
FREUD Anthony Storr
FUNDAMENTALISM Malise Ruthven
FUNGI Nicholas P. Money
THE FUTURE Jennifer M. Gidley
GALAXIES John Gribbin
GALILEO Stillman Drake
GAME THEORY Ken Binmore
GANDHI Bhikhu Parekh
GEOGRAPHY John Matthews and David Herbert
GEOPOLITICS Klaus Dodds
GERMAN LITERATURE Nicholas Boyle
GERMAN PHILOSOPHY Andrew Bowie
GLOBAL CATASTROPHES Bill McGuire
GLOBAL ECONOMIC HISTORY Robert C. Allen
GLOBALIZATION Manfred Steger
GOD John Bowker
GRAVITY Timothy Clifton
THE GREAT DEPRESSION AND THE NEW DEAL Eric Rauchway
HABERMAS James Gordon Finlayson
THE HEBREW BIBLE AS LITERATURE Tod Linafelt
HEGEL Peter Singer
HERODOTUS Jennifer T. Roberts
HIEROGLYPHS Penelope Wilson
HINDUISM Kim Knott
HISTORY John H. Arnold
THE HISTORY OF ASTRONOMY Michael Hoskin
THE HISTORY OF CHEMISTRY William H. Brock
THE HISTORY OF LIFE Michael Benton
THE HISTORY OF MATHEMATICS Jacqueline Stedall
THE HISTORY OF MEDICINE William Bynum
THE HISTORY OF TIME Leofranc Holford-Strevens
HIV AND AIDS Alan Whiteside
HOLLYWOOD Peter Decherney
HUMAN ANATOMY Leslie Klenerman
HUMAN EVOLUTION Bernard Wood
HUMAN RIGHTS Andrew Clapham
THE ICE AGE Jamie Woodward
IDEOLOGY Michael Freeden
INDIAN PHILOSOPHY Sue Hamilton
THE INDUSTRIAL REVOLUTION Robert C. Allen
INFECTIOUS DISEASE Marta L. Wayne and Benjamin M. Bolker
INFINITY Ian Stewart
INFORMATION Luciano Floridi
INNOVATION Mark Dodgson and David Gann
INTELLIGENCE Ian J. Deary
INTERNATIONAL MIGRATION Khalid Koser
INTERNATIONAL RELATIONS Paul Wilkinson
IRAN Ali M. Ansari
ISLAM Malise Ruthven
ISLAMIC HISTORY Adam Silverstein
ISOTOPES Rob Ellam
ITALIAN LITERATURE Peter Hainsworth and David Robey
JESUS Richard Bauckham
JOURNALISM Ian Hargreaves
JUDAISM Norman Solomon
JUNG Anthony Stevens
KABBALAH Joseph Dan
KANT Roger Scruton
KNOWLEDGE Jennifer Nagel
THE KORAN Michael Cook
LATE ANTIQUITY Gillian Clark
LAW Raymond Wacks
THE LAWS OF THERMODYNAMICS Peter Atkins
LEADERSHIP Keith Grint
LEARNING Mark Haselgrove

LEIBNIZ Maria Rosa Antognazza
LIBERALISM Michael Freeden
LIGHT Ian Walmsley
LINGUISTICS Peter Matthews
LITERARY THEORY Jonathan Culler
LOCKE John Dunn
LOGIC Graham Priest
MACHIAVELLI Quentin Skinner
MAGIC Owen Davies
MAGNA CARTA Nicholas Vincent
MAGNETISM Stephen Blundell
MARINE BIOLOGY Philip V. Mladenov
MARTIN LUTHER Scott H. Hendrix
MARTYRDOM Jolyon Mitchell
MARX Peter Singer
MATERIALS Christopher Hall
MATHEMATICS Timothy Gowers
THE MEANING OF LIFE Terry Eagleton
MEASUREMENT David Hand
MEDICAL ETHICS Tony Hope
MEDIEVAL BRITAIN John Gillingham and Ralph A. Griffiths
MEDIEVAL LITERATURE Elaine Treharne
MEDIEVAL PHILOSOPHY John Marenbon
MEMORY Jonathan K. Foster
METAPHYSICS Stephen Mumford
MICROBIOLOGY Nicholas P. Money
MICROECONOMICS Avinash Dixit
MICROSCOPY Terence Allen
THE MIDDLE AGES Miri Rubin
MILITARY JUSTICE Eugene R. Fidell
MINERALS David Vaughan
MODERN ART David Cottington
MODERN CHINA Rana Mitter
MODERN FRANCE Vanessa R. Schwartz
MODERN IRELAND Senia Pašeta
MODERN ITALY Anna Cento Bull
MODERN JAPAN Christopher Goto-Jones
MODERNISM Christopher Butler
MOLECULAR BIOLOGY Aysha Divan and Janice A. Royds
MOLECULES Philip Ball
MOONS David A. Rothery
MOUNTAINS Martin F. Price
MUHAMMAD Jonathan A. C. Brown
MUSIC Nicholas Cook
MYTH Robert A. Segal
THE NAPOLEONIC WARS Mike Rapport
NELSON MANDELA Elleke Boehmer
NEOLIBERALISM Manfred Steger and Ravi Roy
NEWTON Robert Iliffe
NIETZSCHE Michael Tanner
NINETEENTH-CENTURY BRITAIN Christopher Harvie and H. C. G. Matthew
NORTH AMERICAN INDIANS Theda Perdue and Michael D. Green
NORTHERN IRELAND Marc Mulholland
NOTHING Frank Close
NUCLEAR PHYSICS Frank Close
NUMBERS Peter M. Higgins
NUTRITION David A. Bender
THE OLD TESTAMENT Michael D. Coogan
ORGANIC CHEMISTRY Graham Patrick
THE PALESTINIAN-ISRAELI CONFLICT Martin Bunton
PANDEMICS Christian W. McMillen
PARTICLE PHYSICS Frank Close
THE PERIODIC TABLE Eric R. Scerri
PHILOSOPHY Edward Craig
PHILOSOPHY IN THE ISLAMIC WORLD Peter Adamson
PHILOSOPHY OF LAW Raymond Wacks
PHILOSOPHY OF SCIENCE Samir Okasha
PHOTOGRAPHY Steve Edwards
PHYSICAL CHEMISTRY Peter Atkins
PILGRIMAGE Ian Reader
PLAGUE Paul Slack
PLANETS David A. Rothery
PLANTS Timothy Walker
PLATE TECTONICS Peter Molnar
PLATO Julia Annas
POLITICAL PHILOSOPHY David Miller
POLITICS Kenneth Minogue
POPULISM Cas Mudde and Cristóbal Rovira Kaltwasser
POSTCOLONIALISM Robert Young

POSTMODERNISM Christopher Butler
POSTSTRUCTURALISM Catherine Belsey
PREHISTORY Chris Gosden
PRESOCRATIC PHILOSOPHY Catherine Osborne
PRIVACY Raymond Wacks
PSYCHIATRY Tom Burns
PSYCHOLOGY Gillian Butler and Freda McManus
PSYCHOTHERAPY Tom Burns and Eva Burns-Lundgren
PUBLIC ADMINISTRATION Stella Z. Theodoulou and Ravi K. Roy
PUBLIC HEALTH Virginia Berridge
QUANTUM THEORY John Polkinghorne
RACISM Ali Rattansi
REALITY Jan Westerhoff
THE REFORMATION Peter Marshall
RELATIVITY Russell Stannard
RELIGION IN AMERICA Timothy Beal
THE RENAISSANCE Jerry Brotton
RENAISSANCE ART Geraldine A. Johnson
REVOLUTIONS Jack A. Goldstone
RHETORIC Richard Toye
RISK Baruch Fischhoff and John Kadvany
RITUAL Barry Stephenson
RIVERS Nick Middleton
ROBOTICS Alan Winfield
ROMAN BRITAIN Peter Salway
THE ROMAN EMPIRE Christopher Kelly
THE ROMAN REPUBLIC David M. Gwynn
RUSSIAN HISTORY Geoffrey Hosking
RUSSIAN LITERATURE Catriona Kelly
THE RUSSIAN REVOLUTION S. A. Smith
SAVANNAS Peter A. Furley
SCHIZOPHRENIA Chris Frith and Eve Johnstone
SCIENCE AND RELIGION Thomas Dixon
THE SCIENTIFIC REVOLUTION Lawrence M. Principe
SCOTLAND Rab Houston
SEXUALITY Véronique Mottier
SHAKESPEARE'S COMEDIES Bart van Es
SIKHISM Eleanor Nesbitt
SLEEP Steven W. Lockley and Russell G. Foster
SOCIAL AND CULTURAL ANTHROPOLOGY John Monaghan and Peter Just
SOCIAL PSYCHOLOGY Richard J. Crisp
SOCIAL WORK Sally Holland and Jonathan Scourfield
SOCIALISM Michael Newman
SOCIOLOGY Steve Bruce
SOCRATES C. C. W. Taylor
SOUND Mike Goldsmith
THE SOVIET UNION Stephen Lovell
THE SPANISH CIVIL WAR Helen Graham
SPANISH LITERATURE Jo Labanyi
SPORT Mike Cronin
STARS Andrew King
STATISTICS David J. Hand
STUART BRITAIN John Morrill
SYMMETRY Ian Stewart
TAXATION Stephen Smith
TELESCOPES Geoff Cottrell
TERRORISM Charles Townshend
THEOLOGY David F. Ford
TIBETAN BUDDHISM Matthew T. Kapstein
THE TROJAN WAR Eric H. Cline
THE TUDORS John Guy
TWENTIETH-CENTURY BRITAIN Kenneth O. Morgan
THE UNITED NATIONS Jussi M. Hanhimäki
THE U.S. CONGRESS Donald A. Ritchie
THE U.S. SUPREME COURT Linda Greenhouse
THE VIKINGS Julian Richards
VIRUSES Dorothy H. Crawford
VOLTAIRE Nicholas Cronk
WAR AND TECHNOLOGY Alex Roland
WATER John Finney
WILLIAM SHAKESPEARE Stanley Wells
WITCHCRAFT Malcolm Gaskill
THE WORLD TRADE ORGANIZATION Amrita Narlikar
WORLD WAR II Gerhard L. Weinberg

John Haigh

PROBABILITY

A Very Short Introduction

OXFORD
UNIVERSITY PRESS

Great Clarendon Street, Oxford OX2 6DP

Oxford University Press is a department of the University of Oxford.
It furthers the University's objective of excellence in research, scholarship,
and education by publishing worldwide in

Oxford New York

Auckland Cape Town Dar es Salaam Hong Kong Karachi
Kuala Lumpur Madrid Melbourne Mexico City Nairobi
New Delhi Shanghai Taipei Toronto

With offices in

Argentina Austria Brazil Chile Czech Republic France Greece
Guatemala Hungary Italy Japan Poland Portugal Singapore
South Korea Switzerland Thailand Turkey Ukraine Vietnam

Published in the United States
by Oxford University Press Inc., New York

First published 2012

British Library Cataloguing in Publication Data
Data available

Library of Congress Cataloging in Publication Data
Library of Congress Control Number: 2012930333

Typeset by SPI Publisher Services, Pondicherry, India
Printed in Great Britain by
Ashford Colour Press Ltd, Gosport, Hampshire

ISBN: 978-0-19-958848-0

Impression: 13

Contents

Acknowledgements

I thank Leighton Vaughan Williams and Mike Smith for helpful information on horse racing odds, Les Miller for the results of his simulation of roulette spins, Ian McHale for his detailed explanation of how to estimate the winning chances of the 32 teams in the finals of soccer's World Cup, and Derek Robinson for his help with the diagrams. I am particularly grateful to an anonymous referee, whose acute and constructive comments have helped me remove ambiguities, order the material more coherently, and seek to draw out the central ideas of the subject.

I beg forgiveness from those others who recognize themselves as the unacknowledged origin of the information, opinions, or anecdotes in this book, but I have relied on too many sources to give individual credit every time. This is not an academic treatise, with every assertion traceable to its roots, but an attempt to sharpen the reader's appreciation of what the subject of probability is about, how it has developed, and to where it might be applied.

All errors and imperfections that remain are mine alone.

John Haigh
August 2011

List of illustrations

Chapter 1
Fundamentals

The scope of probability

Probability is the formalization of the study of the notion of uncertainty. The effects of blind chance are apparent everywhere. Biologically, we are all a random mixture of the genes of our parents. Catastrophes, like oil spills, volcano eruptions, tsunamis, or earthquakes, and happier events such as winning lottery prizes, randomly and dramatically change peoples' lives.

Many people have a good intuitive understanding of probability. But this understanding can go astray when you have an initial idea about the likelihood of something, but then some new fact, whose relevance is not wholly apparent, is revealed. There are indeed a few notorious 'trick questions', about birthdays, or families with two children, or television game shows with three choices, that seem to have been designed to persuade you that the subject defies common sense. It does not. So long as any hidden assumptions in these questions are flushed out, and taken account of, sensible answers emerge. But probability does require clear thought processes.

The development of its ideas and methods has been driven by its wide applicability. The D-Day invasion of Normandy went ahead in June 1944 only because the probability of favourable weather was deemed sufficiently high. Engineers in the Netherlands must take account of the chances of severe floods when they build the dykes that protect their country from the sea. Is a new medical treatment more likely than present methods to enable a patient to survive for five years? How much you pay to insure your life, car, house, or possessions depends on the chances of an early claim being made. Most decisions you make – what to study at school, who to select as a life partner, where to live, which career to follow – are made under conditions of uncertainty. As Pierre-Simon Laplace wrote in 1814:

> ...the most important questions in life are, for the most part, only problems in probability.

Whenever the phrase 'the probability is...' appears, some assumptions (that may inadvertently have been omitted) are being made. If those assumptions are unwarranted, little reliance should be placed on the claim. I hope that, in this book, these assumptions are clear, either implicitly or explicitly. Before we look at how probability statements can be interpreted, we will describe different ways in which they may arise.

The objective view

The *classical*, or *objective*, view of probability is that often used during games of chance, such as rolling dice, or spinning roulette wheels. There is some list of outcomes: then, either from considerations of symmetry, or because we can find no good reason for one of them to occur rather than another, we take them all as equally likely. So we just count the number of outcomes, and give them all the same probability. Then the probability of any event in the experiment is taken as the *proportion* of outcomes that favour it.

For example, when a coin is thrown twice, the four possible Head/Tail outcomes are HH, HT, TH, TT. With a *fair* coin, H or T will be equally likely each time, so none of those four outcomes should be more or less likely than any of the others, each should have probability 1/4. Three of them contain Heads at least once, so the probability of the event that Heads appears at all is 3/4.

There are 1,326 ways of dealing a hand of two cards. (Take my word for it.) If the deck has been well shuffled, we take all these hands as equally likely. And 64 of them consist of an Ace and a 'ten-card' (i.e. Ten, Jack, Queen, or King), so we conclude that the probability of being dealt such a hand – 'Blackjack' – is 64/1326, just under 5%.

So far as probability considerations are concerned, both these examples could be reformulated in terms of choosing one ball from a bag of identical balls. The first bag would have four balls, three of them Red, the second 1,326 balls, 64 of them Red. Indeed, every example in this objective approach to probability is essentially identical to some problem about selecting one ball from some bag or urn (which perhaps explains the plethora of such exercises in student textbooks).

I emphasize that it is not enough to count the number of possible outcomes, and how many of them favour the event in question. There must also be no cogent reason for any outcome to be more or less likely than some other. Otherwise, you could fall into the trap of believing that your chance of winning the Jackpot in a Lottery is 50%, on the grounds that there are just two alternatives, either you win or you do not!

Experimental evidence – frequencies

We hope that dice used in household games like Monopoly, or casino games like Craps, will show each of their six faces equally often. But if a die is made from non-uniform material, or its width, breadth, and height differ, it is not sensible to assume that all outcomes are equally likely. Over a series of throws made under the same conditions, the frequency of any face will fluctuate, but will eventually settle down close to some particular value.

You do not find that 20% of the first thousand throws are Sixes, and then the proportion among the next thousand throws leaps up to 60%. In these repeatable experiments, the outcomes may not be equally likely, but each of them has a propensity to occur at some characteristic frequency, and a *frequentist* takes this value as the probability of that outcome.

Perhaps we get 170 Sixes in the first thousand throws of an imperfect die, then 181 Sixes in the next thousand, and so on. We can never deduce an *exact* value for the probability of a Six from these experiments, but the data lead to estimates, and the more data that are collected, the better we expect the estimate to be. The fact that we cannot know the exact probability does not deny its existence.

If I draw one card from a well-shuffled pack, there seems no reason for one suit to be favoured over any other. Each suit would have objective probability of 1/4. And if I return the card, reshuffle, and perform this task one hundred times, I expect each suit to arise about equally often, in this case about twenty-five times. Similarly, with ordinary dice where all six outcomes are intended to be equally likely, the chance of a Five on any throw is objectively taken as one-sixth: and over six hundred throws, we expect a Five on about one hundred occasions.

When experiments with equally likely outcomes are repeated often, the relative *frequency* of any particular outcome is expected to be a close match to its probability, as calculated objectively. A fair coin seldom gives exactly fifty Heads in one hundred throws, but intuition does not tell you how close to that ideal you should reasonably expect.

Frequency ideas are applied more widely than to repetitions of the same experiment under identical conditions. Will some imminent birth be male or female? With no specific information about the family in question, turn to data gathered from many countries and cultures over a long period. There is a consistent pattern that, for every 49 female births, there are 51 males. On the basis that there is nothing to pick out this birth from all others that are taking place, a frequentist will put the probability of a boy at 51%.

Some experiments on a heroic scale have been conducted. In 1894, the zoologist Raphael Weldon reported the results of more than twenty-six thousand throws of a set of a dozen dice. His data were *not* consistent with the idea that all six faces were equally likely, as the numbers five and six occurred rather too often. His dice had small holes drilled in each face to identify its score, and the faces for five and six are opposite two and one respectively. The centres of gravity of these dice will be closer to the faces with small numbers, giving a plausible explanation for the observed excess.

About seventy years later, Willard Longcor, a meticulous man with time on his hands, offered his services to top Harvard statistician Frederick Mosteller. Under Mosteller's guidance, Longcor collected over two hundred dice, and threw each of them twenty thousand times, recording the outcome simply as even or odd – over four million data values. To make the conditions as near as possible identical, he used a carpeted desk-top, with a raised step to bounce the dice off. For cheap dice like those used by Weldon, there was a small but distinct bias towards too many even numbers – again, not totally unexpected because of the drilling. However, with the high quality precision dice as used in Las Vegas casinos, where the pips are either lightly painted or are extremely thin discs, no such bias was found. Frequencies with those dice were consistent with the classical view of equally likely outcomes.

Blackjack expert Peter Griffin noted wryly that, for a sequence of 1,820 hands he played in Las Vegas, the dealer's upcard was either a Ten-card or an Ace on 770 occasions. The objective chance of receiving one of those favourable cards is 5/13, so Griffin wondered whether or not he had been cheated – random chance would give the dealer these good cards only 700 times on average.

In 2002/3, 6,202 children under five years old were admitted with suspected pneumonia to hospitals in Malawi, and 523 died, a fatality rate of 8.4%. Provided that there were no special circumstances making this period atypical, a frequentist would conclude that the probability of death when a young Malawi child catches pneumonia is about 8–9%. From an objective perspective, making general statements about the chance of death among young Malawi children with pneumonia would be speculation, albeit based on evidence: but all that can be said for certain is that if one of *those particular* 6,202 children were selected at random, the chance that child died was 8.4%.

The relationship between frequency data and objective probabilities will be further explored later.

The subjective interpretation

Bruno de Finetti, one of the most influential thinkers in the field, wrote

PROBABILITY DOES NOT EXIST

As Professor of the Theory of Probability, he was not dismissing his subject as a mirage, rather he rejected *absolute* claims such as 'The probability of Heads is one half'. To him, every statement involving a probability is just an expression of opinion, based on one's own experience and knowledge, and perhaps changing when more information arrives.

Consider the five assertions:

> The England cricket captain will win the toss in England's next Test Match;
>
> Whoever wins the Oscar for best actor next year will also win it the year after;

No person born in Oslo has yet won an Olympic fencing gold medal;
Richard III was responsible for the death of the Princes in the Tower;
Al Gore would have been elected US President in 2000 if Ralph Nader had not stood as a candidate.

To each of them, we can offer our *degree of belief*, or *personal probability*, or *subjective probability*. This will be some non-negative number, not greater than 1: equivalently, it is a percentage between 0% and 100%, inclusive.

Zero and one represent, respectively, the two extremes of *impossible*, and *certain*. I am certain that the soccer World Cup will be hosted by an African nation again during the present century. I think it is impossible for someone under twenty years of age to win a Nobel Prize.

Assessing subjective probabilities

The five assertions above have different natures, and we have different kinds of evidence about them. For the first, we might appeal to symmetry between Heads and Tails. For the second, we have the history of the Oscars since 1929 to guide us. In both these cases, the truth or otherwise of the statement will become known within a finite time. The third is either true or false, and could be established now by a thorough trawl of Olympic records. The fourth is also either true or false, but we will never know which. We cannot rerun history to ascertain the truth or otherwise of the fifth claim.

Specific examples later will illustrate how subjective probabilities have been assessed. Aside from those arguments, there are at least three distinct general approaches. One is as the *fair price* for a bet that the event will occur. But this does not work for everybody: some people have principled objections to betting, others are unwilling to contemplate actions that might ever lead to a loss. And even for those who do feel comfortable with betting, their fair price might differ according as to which side of the bet they were on.

A second way to assess your degree of belief in an event uses the objective approach. Which offer would you prefer: to receive £5 if the event occurs, or to receive £5 if you correctly guess the colour, Red or Black, of the top card in a well-shuffled deck? If you prefer the latter, your degree of belief in the event is below 50%.

Suppose that is the case. Now compare the prospect of receiving £5 if the event occurs, or getting it if you correctly guess the *suit* of a randomly drawn card. The latter should occur 25% of the time, so your preference here will tell you whether your degree of belief is below 25%, or is between 25% and 50%.

More elaborate comparisons along these lines let you home in on a situation where you cannot say on which side your preference lies. Your degree of belief in the event will then be close to the objective probability of the corresponding card selection. Rather than use a deck of 52 cards, with its awkward fractions, you might think of an urn containing 20, or maybe 100, identical balls with which to specify the alternative events.

Give your answers with appropriate precision. Tennis players John Isner and Nicolas Mahut played the longest match in Wimbledon history in 2010; via counting, the chance that they would be drawn together again the next year (it happened!) is precisely 2 in 285, perhaps better rounded to 'a little under 1%'. But it was absurd of *Star Trek*'s Mr Spock to tell Kirk that the odds against their escape in one episode were 'approximately 7,824.7 to 1'.

For a third method, think of a modest sum of money, not so small that you are totally indifferent to it (say, one penny), nor so large that possessing it would make a dramatic difference to your circumstances (£1 million to most people, rather bigger for Bill Gates). For me, £10 fits the bill – call this *unit amount.*

Now suppose that, somehow or other, the truth or falsity of the event will be revealed tomorrow: and you will receive this unit amount if it is true, but zero if it is false. But rather than wait for tomorrow, you could receive a definite proportion p of this unit amount today. (Getting the money today or tomorrow makes no difference to you.)

If p is tiny, you are likely to reject the offer, and will prefer to wait; if it is close to unity, you may well accept that definite amount. But there will be some intermediate value of p where you are indifferent between taking this offer, and waiting for the outcome to be revealed. This p is your degree of belief about this statement or event in question.

I offer my own subjective answers for the five assertions above. I can think of no sensible reason why one side should be more or less likely to win a cricket toss than the other, so my first figure is 50%. Looking at Oscar history, not only for actors but also the other categories, the award has occasionally been repeated in successive years: perhaps there are more candidates these days, leading me to suggest 3%, or lower. Norwegians are not noted for fencing, but we have épée, foil, sabre, and the sport has appeared in all the Summer Games from 1896. Some native of Oslo might have won sometime, but I strongly doubt it – my figure here is about 95%. Prejudice in favour of the White Rose county, rather than objective evidence, leads me to suggest 10% for the fourth claim. For the fifth claim, considering the votes in each State, and thinking of a plausible division of the votes Nader received, guide me towards 20%.

Pause a while, and make your own suggestions for these five claims. The better you are at assessing probabilities when matters are uncertain, the more likely you are to be happy with the decisions you make in life.

Odds

Whether we use the classical approach, or frequencies, or degrees of belief, the term *odds* is often used when describing probabilities. We might say that the odds of obtaining a Six with a fair die are 'five to one against' – for every time we get a Six in a sequence of throws, we expect not to do so five times. If an outcome is expected to be more likely than not, such as the higher ranked player winning a tennis match, that event is said to be *odds on*.

There is an exact correspondence between probabilities and odds, and we can switch easily between them. Thinking of frequencies can help. If the probability is 20%, or one fifth, we expect the event to occur in one occasion out of five, so the odds are 'four to one against'. For a probability of 75%, we expect it to occur three times out of four, giving odds of 'three to one on'. And if the odds are stated as six to five against, this indicates that for each five times the event happens, it fails to do so six times, so its probability is 5/11.

You do not have to stick to whole numbers. The probability that the top card in a well-shuffled deck is either a King or a Queen will be taken as 2/13. This could be quoted as 'eleven to two against' or, equally accurately, 'Five point five to one against'. Use whichever you like.

Although the phrase 'the odds are one to one' is never used, it would make perfect sense. It indicates that an event is expected to happen just as often as not, so its probability is one half. Instead, with a straight face, we say 'the odds are evens'.

Issues to resolve

There are no important disagreements about how to work with probabilities, but adherents of the three approaches we have described may deduce their values in different ways. Each perspective has its uses. In seeking to understand how the subject works, we will appeal to whichever viewpoint appears appropriate.

The objective approach is limited to circumstances having finitely many outcomes, all judged equally likely. But no coin or die is perfectly symmetrical, and on what basis can we dismiss its imperfections as irrelevant? Can we even be sure that we agree on the number of possible outcomes? For example, suppose we are told that an urn contains two balls, either both White, both Black, or one of each colour. Should we argue we have *three* equally likely cases, or that there are really *four* equally likely cases, arising when the balls were inserted, in order, either as WW, WB, BW or BB? These different outlooks would give different answers for the chance that both balls are Black. Or suppose you reach a road junction with three possible exits, two of them leading to Newtown, the third to Seaport: making a 'random choice', is the chance that you aim for Seaport one third (one exit in three), or one half (one of two destinations)?

A frequentist seeks to deal with circumstances that are repeatable indefinitely often under identical conditions. The number of outcomes need not be finite – think of tossing the same coin until Heads appear three times in succession, or selecting a random point on a stick. But, however much care we take, the experimental conditions cannot be *absolutely* identical, and any limiting value can only be estimated. How should the error in this estimate be described? Claiming that the probability is at least 99% that the error is under 2% requires a circular argument – we need to know what probability is, in order to define it!

For questions such as the probability that one country invades another, or the chance that a particular heart transplant is successful, the circumstances arise once only, and the alternatives cannot be reduced to a finite list of equally likely cases. The objective and frequency approaches are silent on these matters. A subjective approach is required.

A subjectivist must ensure that her beliefs are consistent with each other. For example, in the UK National Lottery, a machine selects six numbers from the list {1, 2, 3,...,49}, and Susie may be content to take all 14 million or so possible selections as equally likely. Then, when asked which is more likely,

(a) that no number drawn exceeds 44, or
(b) that those drawn do not include two consecutive numbers,

she may, after a little thought, come down on one side or the other. But if she selects *either* of these events as more likely than the other, she will be guilty of inconsistency, as proper counting shows that they can occur in exactly the same number of ways! Nothing in the subjective approach specifies how such an inconsistency should be resolved, merely that it must be.

Because we wish to think about probabilities in circumstances wider than when there are finitely many equally likely choices, and in circumstances that cannot be repeated indefinitely often, we will take the subjective approach as the default option. But we are likely to hold more firmly to our opinions when they are backed up by either an objective, or by a frequency, argument.

Interpretations

Using the 'balls in a bag' viewpoint, the probability of some event is taken as the proportion of Red balls in the bag. So a value of zero can occur only if there are no Red balls, in which case the event will never happen. Similarly, a probability of unity corresponds to every ball being Red, so here the event occurs every time. These values, zero and unity, are the only ones that can be conclusively proved *wrong* by experimental evidence: if the event happens, its probability cannot be zero, if it fails to happen, its probability cannot be unity. And this is true for the frequency, or subjective approaches also. So suppose the probability has some intermediate value, say 3/4.

We first dispose of one finicky point. No matter how well a roulette wheel has been engineered, it is physically impossible that all the numbered slots have *exactly* the same chance. What the casino requires is that the chances are close enough to the ideal that it is inconceivable that any number could be picked out as more or less likely than another. Similar remarks apply to dice, coins, or cards. So statements like 'the probability is 3/4' will mean that the probability is close enough to 3/4 for all practical purposes. Otherwise, a pedant might smugly tell you that he *knows* that the probability is not 3/4, without fear of contradiction.

In the context of repeatable experiments, what do we expect to follow from the claim: 'The probability of a Red ball is 3/4'? Emphatically, we do not expect that if we conduct this experiment four times (replacing the ball drawn on each occasion), we shall draw a Red ball in precisely three of them. It is possible that four repetitions throw up no Reds at all, or even that Red happens every time. But over a long series of repetitions, we do expect the overall frequency of Red to be close to 3/4.

There are no black/white answers to what constitutes a long series of experiments, nor to how close to 3/4 is acceptable. If I obtained Red only 20 times in the first 40 repetitions, I would have very strong doubts about a claim that the probability was 3/4; but those doubts would be largely assuaged if the next 40 repetitions gave 28 Reds. Believing or disbelieving this claim can be a provisional position for quite some time. Assuming the experimental conditions do remain unchanged throughout, use *all* the data collected to reach a decision – short runs can mislead.

I will offer some guidelines, and justify them later. Take the case when we make one hundred repetitions, and the supposed probability is some middling value, near one half. Compute the *difference* between this figure and the actual frequency from data: if this difference exceeds 0.1, I would have some doubts about the claim, and if it exceeded 0.15, I would have strong doubts. With a thousand repetitions rather than a hundred, I expect closer agreement, so replace those numbers by 0.03 and 0.05. If the supposed probability is closer to zero or unity, say 10% or 90%, I would also require better agreement. It can be much easier to be convinced, on the basis of repeated experiments, that a particular probability is *not* some alleged value.

What about a subjective assessment, such as that the probability of rain tomorrow is 60%? We cannot recreate today's weather conditions hundreds of times, and check how often it rains. This 'experiment' can be conducted once only. But we might test the claim by looking at the process that led to it being made. Forecasters use models of weather patterns to reach their conclusions, and even if the figure on their computer screen is 31.067%, they sensibly offer round figures. You hear 'The chance of rain is about 30%'. So now you can collect data for different days, and look at the empirical evidence – in how many of the 83 days last year when the chance of rain was put at 30% did it actually rain? So long as that proportion was reasonably close to 30%, your belief in the method is reinforced, so accepting the figure given for 'tomorrow' is a rational response.

Probability is the key to making decisions under conditions of uncertainty. If you honestly believe that the probability of a particular event or statement is unity, you should act as though it will definitely occur; and if your honest belief is that the probability is zero, act as though it cannot occur.

If you think the probability is some value between zero and one, act as though you expect it to occur that proportion of the time.

For example, if your judgement is that the probability is 60%, imagine that you will face this situation a hundred times, in sixty of which (but you have no idea which sixty) this event will happen, and forty times it will not. Swallow hard, and decide on your action, taking into account this balance. Had you judged the probability to be 80%, so that you expect the event to happen rather more often, your action might well be different.

As Bishop Joseph Butler wrote in his 1736 *Analogy of Religion*, 'To us, probability is the very guide to life'.

Chapter 2
The workings of probability

As well as the subjective, objective, and frequentist approaches to probability, there are other standpoints. For example, should one always insist on associating a probability with a number? Might it be enough to say that one probability was greater, or one degree of belief was more intense, than another? And should we necessarily offer an initial set of axioms – self-evident truths – on which to erect a theory?

Many distinguished writers have felt it useful to have two separate approaches, one for degrees of belief and one for objective probabilities. Both would have the same rules of logic, free from contradictions, but how values of probabilities were arrived at, and how they are interpreted, could differ. Any theory should be *consistent* with the classical view, based on repeatable experiments with equally likely outcomes. So we will focus on that case, seeking any rules that the notion of probability must obey.

The Addition Law

Deal one card from a well-shuffled pack. We take all cards as equally likely, so the probability of any event, such as obtaining a Club, or a Spade, or an Ace is found by calculating the proportion of all possible outcomes that lead to those events. How might we find the probability that *either of two* such events occur?

If those events have no outcomes in common, we say that they are *mutually exclusive*, or *disjoint*. The events 'Get a Spade' and 'Get a Club' are disjoint, but the events 'Get a Spade' and 'Get an Ace' are not disjoint, as the Ace of Spades belongs to both. When two events are mutually exclusive, then the total number of outcomes that lead to either event is just the sum of the numbers for each event separately, so we have a simple result: *whenever two events are mutually exclusive*,

the probability that at least one occurs is the sum of their individual probabilities.

This is the *Addition Law*. It plainly holds in all experiments where we would take the classical view: using the balls in a bag analogy, it is the same as saying that the total number of balls that are either Red or Blue is the sum of the number of Red balls and the number of Blue balls. And in any repeatable experiment, such as rolling dice, or spinning roulette wheels, the sum of the individual frequencies of two disjoint events is inevitably the frequency that at least one of them occurs. So frequentists accept the Addition Law too.

Also a subjectivist accepts this Law. For otherwise, there would be two disjoint events, call them A and B, where it did not hold. In that case, the subjectivist could be confronted by three bets: one about A, one about B, and one about either A or B, and would accept each bet on its own as fair. *But* he could be guaranteed to lose money if all three bets were struck! The Addition Law forbids this inconsistency.

This Addition Law extends to a collection of many events, provided no two of them have any outcomes in common – they are *pairwise disjoint*. The probability that at least one among even millions of pairwise disjoint events occurs is just the sum of their individual probabilities. But suppose the number of outcomes is no longer finite: for example, tossing an ordinary coin repeatedly until Heads appears for the first time.

The possible outcomes of this experiment are the unending list {1, 2, 3, 4,....}, each with its own non-zero probability. What is the chance that we take some *even* number of throws to get a Head? That event happens when any of the outcomes {2, 4, 6, 8,....} happen. Could we compute its probability by adding up the corresponding individual probabilities?

There is no mathematical difficulty in doing this adding up, but that action falls outside the scope of the classical view of probability, which deals only with a *finite* list of outcomes. There is no consensus as to whether the Addition Law for such an *unending* list should be part of the workings of probability. In favour of including it, we may be able to find the probabilities of a wider class of events than without it; against inclusion, as it is not part of the classical theory, we should be cautious about taking steps that might have hidden pitfalls. There is no right or wrong answer.

I'm a pragmatist. I am content to extend the use of the Addition Law in this way, and I have never felt uncomfortable with the results of doing so. This position is a standard part of the account given in most books used to teach the subject at university. But de Finetti took the cautious view to avoid making this extension, and others have felt the same way.

The Multiplication Law

If you toss an ordinary coin, you will expect to guess Heads or Tails correctly half the time. If you shuffle a deck of cards, and predict whether the top card is Red or Black, you also expect to be correct half the time. When you guess both a coin toss and a card colour, how likely are you to get *both* correct?

Think of conducting this double experiment a hundred times. You expect to guess the coin correctly about fifty times, and when you do so, you expect to go on to guess the card colour half the time.

That suggests you expect to get both right on about twenty-five occasions, and it looks sensible to offer 25%, or 1/4, as the chance of being right both times. For these experiments, the chance of being correct both times is found just by multiplying their individual chances.

Ten balls of the same size and composition are labelled with the numbers zero up to nine, and one of them is selected completely at random. So it is equally likely to show a Low number (zero to four), or a High number (five to nine). Five of these numbers are coloured Green, the rest are Blue, so Green and Blue are also equally likely. Trying to guess the colour, or whether it is Low or High, we have a 50% chance either time. What about the chance that the ball we draw is both Low and Green?

The argument above with the coin and the cards suggests 1/4 as the answer, but a moment's thought shows this cannot be correct. With ten balls, it is impossible that 1/4 of them (two and a half!) will be both Low and Green! The correct answer depends on *which* numbers are coloured Green, and which Blue. So suppose numbers one to five are Green, the rest are Blue.

In that case, four of the ten numbers (one, two, three, and four) are both Low and Green, so the chance is 0.4. But, as we did with the first problem, we can also use a two-stage process: in one hundred repetitions of this experiment, we expect to get a Low number fifty times. Four of the five Low numbers are Green, so when we did get a Low number, we expect it to be Green 4/5 of the time. Overall, we expect a Low Green number forty times, leading again to the answer 0.4.

With the coin and cards, the outcome of the coin toss has no bearing on the card drawn. We do not change our minds about the chance of a Red card when told whether the coin falls Heads – the *conditional* probability of the second event, *given* the first, is just its ordinary probability. When this happens, the two events are said to be *independent*, and the chance that both occur is the product of their individual probabilities.

With the ten balls, the chance both events occur also arises as a product, in which the first component is also the probability of one event (Low number), and the second is the *conditional* probability of Green when this happens. So really the two calculations are identical in form, the only difference is whether the outcome of the first event affects the chance of the second. Each time, we have used the *Multiplication Law*, which says:

> the probability that both of two events occur is the probability of the first, multiplied by the probability of the second, conditional on the first happening.

Independence

We used the term 'independent' to describe the circumstances when the occurrence of the first event does not change our assessment of the chance of the second. Suppose this holds, but we learn that the *second* event has happened; might this affect our assessment of the chance of the first?

No. Whenever the fact that one event has or has not occurred makes no difference to the chances of another event, it turns out that whether or not this second event occurs makes no difference to the chances of the first. Two events are independent when the occurrence or non-occurrence of either makes no difference to the probability of the other. To find the chance both occur, multiply their individual chances.

Events that have no bearing at all on each other, like rain today in Tunis and the gender of the next birth in Paris, are surely independent. But sometimes independence is not obvious. Using an ordinary fair die, consider the events 'Get an even number' and

'Get a multiple of three', with respective chances one-half and one-third. The only way both occur is when we get a Six, having probability one-sixth. And since multiplying one-half and one-third gives one-sixth, those two events *are* independent. The chance of getting an even number does not change if we are told whether or not a multiple of three occurs (and vice versa).

Now consider the same problem when you have an eight-sided fair die, or a ten-sided fair die, with the sides labelled one to eight or one to ten respectively. Do the arithmetic: you should find that the two events *are* independent in one of these cases, but *not* independent in the other. Intuition about independence is useful, but not always enough.

Assuming two factors to be independent, when they are not, is one of the most common mistakes made in assessing probabilities. Suppose that, in a university's graduate school, half the students are female, and one in five study engineering. Select one student at random: the probability this student is female can be taken as one-half, the probability the student studies engineering will be one-fifth. However, you will find that the probability the student is a female engineer is much less than their product, one-tenth.

Events with overlap

The Addition Law shows how to find the chance of at least one of two events, provided they are disjoint. What if they are not disjoint? For example, in drawing one card at random, what is the chance it is either a Spade or an Ace? The Ace of Spades falls in both categories, so if we just added the respective probabilities, we would count that card twice. To correct for the outcomes that would be double-counted and find the probability that *at least one* of two events occurs,

> add their individual chances, then subtract the chance both occur.

Of course, if the two events are disjoint, it is impossible that both happen, so this extra term has value zero, and we are back to the original Addition Law.

Let's see this notion in action in the two earlier examples. With the coin and Red/Black card, the chance we get at least one guess right comes from the arithmetic 1/2+1/2-1/4, which is 3/4. In the other example, the chance a randomly chosen numbered ball is either Low or Green is 1/2+1/2-0.4 = 0.6.

And the chance of drawing either a Spade or an Ace arises as 13/52+4/52-1/52 = 16/52, confirmed by noting that exactly 16 of the 52 cards satisfy this condition.

This last calculation should warn you against making early arithmetic simplifications. Yes, 13/52 is the same as 1/4, and 4/52 is the same as 1/13, but to add 1/4 and 1/13, you are better off with their original fractions. And it is seldom helpful to re-write a friendly fraction like 5/13 as its ugly decimal approximation 0.38461538...

More than two events

A collection of many events is described as independent whenever knowing whether or not some of them occur makes no difference to the probabilities of any of the others. In this case, the Multiplication Law means that, whatever selection of events we make from this collection, the probability that all of them occur is just the product of their individual probabilities.

But how might we find the probability that three or more events all occur, when they are *not* independent? For example, in whist or bridge, the cards are randomly shuffled and shared equally among four players. How likely is it that they all receive exactly one Ace?

Consider the four separate events: Anne gets exactly one Ace; Brian gets one Ace; Colin has one Ace; Debby gets one Ace. Plainly, these four events are not independent, as if any three of them happen, the other is certain. We will find the probability they all occur via a *three-stage* process.

First, we find the probability that Anne has exactly one Ace. Assuming all possible ways of dealing the cards are equally likely, we have an exercise in counting: count the total number of possible deals, and then count in how many of them she gets exactly one Ace. Believe me, the chance works out as just under 44%.

Assume Anne has just one Ace (and hence twelve non-Aces). That leaves three Aces and thirty-six non-Aces for the other players, and Brian gets thirteen of them, chosen at random. A similar counting exercise on this smaller pack shows that the chance he would get exactly one Ace is just over 46%. The Multiplication Law then tells us that the chance of both events, i.e. that both Anne and Brian have exactly one Ace, is the product of these two values, just over 20%.

So now assume Anne and Brian each have exactly one Ace. Then Colin receives thirteen cards at random from the two Aces and twenty-four non-Aces that remain: the chance that he gets exactly one Ace is found to be 52%.

The final step is to use the Multiplication Law once more, to combine these last two calculations: the chance that Anne and Brian, and then also Colin, all have exactly one Ace is a little over 10%. If this happens, Debby inevitably has the final Ace, so we have found the answer we seek.

This answer itself is of no real consequence – although the deal is totally random, the most equitable outcome for the Aces is rather unlikely – but the *method* used is universal. To find the chance that every event in a collection occurs, break things down into stages. Find the chance for one event; then, *assuming* this event occurs, find the chance of a second; now *assuming both* of these occur, find the chance of a third; then *assuming all these three* events happen, find the chance of a fourth – and so on. Finally, multiply all these quantities together.

Where else might we have to follow this path? Suppose my journey has three stages, and I can assess their separate chances of having no delay: however, all the stages will be affected by the weather, and delay on one stage will change the chance of delay elsewhere. In manufacturing industry, the safety of a piece of equipment will rely on several components which do not operate independently – some may use the same water supply, others may have been inadequately tested by the same unreliable employee. With a medical procedure, whether or not the things that can go wrong are independent of each other can make a huge difference to the overall chance that all turns out well.

If events are independent, then the chance they all occur is just the product of their individual chances. But we are seldom lucky enough for this condition to hold: a stage-by-stage assessment, with probabilities changing as the work progresses, is the norm.

What about the chance that *at least one* of three or more events occurs? The Addition Law does extend to this case, but as the expression is cumbersome, I will not write it down. Its recipe follows the same path as described when using the Multiplication Law for the chance that all of many events occur: take it one step at a time.

Using the word *independent* when *disjoint* is meant, and vice versa, are common errors. The example of choosing one card at random can help you see how to avoid it. Here, the events 'Get a Spade' and 'Get a Club' are disjoint, but far from independent, as if either occurs, the other cannot, so the chance both occur is zero! Also 'Get a Spade' and 'Get an Ace' are independent (yes?), but plainly not disjoint.

Remember: the Addition Law is used to find the chance of at least one event, the Multiplication Law is used to give the chance they all occur.

It is sometimes said that counting really goes one, two, infinity. This aphorism carries the truth that if we can make the step from dealing with one case to dealing with two cases, then subsequent steps to three, four, five, etc. cases are trivial in comparison. This surely applies to both the Addition and Multiplication Laws.

A neat trick

Any event either happens, or it does not. The total probability is split between the event happening, and it failing to happen. So if we can find the chance an event does *not* happen, we can deduce the chance it does occur by subtraction from 100%.

To illustrate, let's find the chance of at least one Six when an ordinary fair die is rolled twice. Any outcome is a pair of numbers showing the scores of the first roll, then the second, e.g. (5, 2) or (4, 4), and we take all such outcomes as equally likely. Each roll has six possible results, leading to 6x6=36 outcomes altogether. Our event does *not* happen when neither die shows a Six, for which there are 5x5=25 outcomes. The chance of no Sixes is 25/36, so the chance of at least one Six is 11/36, a bit less than one in three.

This leads on to a junior version of a gambling problem solved by Blaise Pascal and Pierre de Fermat in 1654. How often must we roll a die to make it *more likely than not* that we get at least one Six, i.e. that the chance of getting a Six is more than one half? We have just seen that two rolls are not enough.

Each extra roll increases the number of possible outcomes by a factor of six, while the number of outcomes without a Six gets multiplied by five. So a third throw generates 216 outcomes in all, and 125 of them – over half – contain no Six; three rolls are not enough either. However, four rolls give 1,296 outcomes, and only 625 of them have no Sixes – fewer than half. That leaves more outcomes that include a Six than outcomes with no Sixes, so now a Six is more likely than not. Four rolls suffice.

The actual game analysed by Pascal and Fermat involved rolling not one die, but two dice together; and asking how often this needs to be done to make it more likely than not that a *Double-Six* turns up at least once. The method of solution is the same, but the raw arithmetic is formidable. Today, we can quickly reach the answer with a microcomputer or a pocket calculator, while logarithms and slide rules had conveniently just become available in the 17th century. With up to 24 rolls, it is more likely than not that no Double-Six appears, but a 25th roll tips the balance the other way.

Most problems of the form 'Find the chance of at least one of these events happening' are best solved in this manner: work out the chance that none of them arises, and then subtract from unity.

Chapter 3
Historical sketch

Beginnings

A game popular in Florence around 1600 rested on the total score from three ordinary dice. The scores of Three, when all dice scored one, and Eighteen when they all scored six, arose rarely, with most scores near the middle of the range. You should check that there are six different ways of scoring Nine (e.g. 6+2+1, 5+2+2, etc.), and also six ways of scoring Ten. It was commonly believed that this 'ought' to make totals of Nine or Ten equally frequent, but players noticed that, over a period of time, the total of Ten occurred appreciably more often than Nine. They asked Galileo for an explanation.

Galileo pointed out that their method of counting was flawed. Colour the dice as Red, Green, and Blue, and list the outcomes in that order. To score Nine from 3+3+3 requires all three dice to show the same value, and that can happen in one way only, (3,3,3). But the 5+2+2 combination could arise as (5,2,2), (2,5,2), or (2,2,5), so this combination will tend to arise three times as often as the former; and 6+2+1 arises via (6,2,1), (6,1,2), (2,6,1), (2,1,6), (1,6,2), and (1,2,6), so this combination has six ways to occur. A valid approach to how often we can get the different totals takes this factor into account, and does indeed lead to more ways of obtaining Ten than Nine. The Florentine gamblers learned a vital lesson in probability – you must learn to count *properly*.

In the summer of 1654, Pascal (in Paris) and Fermat (in Toulouse) had an exchange of letters on the *problem of points*. Suppose Smith and Jones agree to play a series of contests, the victor being the first to win three games; unfortunately, fate intervenes, and the contest must end when Smith leads Jones by 2-1. How should the prize be split?

Such questions had been aired for at least 150 years without a satisfactory answer, but Pascal and Fermat independently found a recipe that, for any target score, and any score when the contest was abandoned, would divide the prize *fairly* between them. They took different approaches, but reached the same conclusion, and each showered praise on the other for his brilliance. For the specific problem stated, the split should be in the ratio 3:1, with Smith getting 3/4 of the prize, Jones 1/4.

The essence of their solution was to suppose that both players were equally likely to win any future game. They counted how many of the possible outcomes of these hypothetical games would give overall victory to either player, and proposed dividing the prize in the ratio of these two numbers. In different language, the prize should be split as the ratio of the two *probabilities* of either player winning the series, assuming they were evenly matched in future games. The systematic study of probability had begun.

This issue was settled via the objective approach to probability, but Pascal also thought more widely. He suggested a wager about the existence of God. 'God is, or is not. Reason cannot answer. A game is on at the other end of an infinite distance, and Heads or Tails is going to turn up. Which way will you bet?'

He argued that if God exists, the difference between belief and unbelief is that between attaining infinite happiness in heaven, or eternal damnation in Hell. If God does not exist, belief or unbelief lead to only minor differences in earthly experience. Thus an agnostic should lean strongly to belief in God.

In this game, the values of the chances of 'Heads' or 'Tails' are personal choices, not derivable from symmetry or counting arguments. Thus Pascal was a pioneer in the subjective approach to probability too.

The Swiss Family Bernoulli

During the 17th and 18th centuries, members of the Bernoulli family from Basle made significant advances in mathematics, including probability. Rivalry was a spur: one of them would pose challenges, another would respond, the originator of the challenge would claim to find flaws in the supposed solution, and so on.

Games of chance inspired much of the early interest in the workings of probability. In these games, be it rolling dice, dealing cards, or tossing coins, some 'experiment' is carried out repeatedly under essentially the same conditions. The natural question, raised earlier, is: how does the *observed* frequency of an outcome relate to its *objective* probability?

Jacob Bernoulli gave an answer in his posthumously published *The Art of Conjecturing* (1713), nicely illustrated by his example. Suppose 60% of the balls in an urn are White, the rest are Black, and one ball is drawn at random. That ball is replaced, and the experiment repeated many times. Bernoulli showed that, so long as at least 25,550 drawings are made, for every time the proportion of White balls falls *outside* the range from 58% to 62%, it will fall *inside* that range at least one thousand times. Informally, the observed frequency of White balls is, in the long run, overwhelmingly likely to be close to its objective probability.

A similar analysis applies to any experiment that can be repeated indefinitely under identical conditions, where the result of one experiment has no effect on the others. Each time, certain outcomes denote Success, and their objective probability is some fixed value p. (This notion now carries the label *Bernoulli trials.*) Take any interval, as small as you like, around the value p – plus or minus 2%, plus or minus 0.1%, it matters not. Also, say how much more often you want the running frequency of Successes to be inside this interval, rather than outside it – a hundred times as often, a million times, whatever. Bernoulli's methods show that *any* such demand can always be met, provided the experiment is repeated often enough. The observed frequency will be as close to the objective probability as you like, given enough data. This assertion is known as the *Law of Large Numbers*.

The family's fame was honoured in 1975 by the name choice 'The Bernoulli Society' for an international society whose main purpose is to foster advances in the study of probability and mathematical statistics.

Abraham de Moivre

De Moivre settled in England as a Huguenot refugee, and made a living from chess and from his knowledge of probability. Isaac Newton, then over 50 years old and with many calls on his time, deflected enquiries about mathematics with the words 'Go to Mr de Moivre, he knows these things better than I do.' De Moivre's *Doctrine of Chances* appeared in English in 1718, and its second edition, in 1738, contained a major advance on Bernoulli's work. To appreciate what he did, consider something specific: if a fair die is rolled 1,000 times, how far from the average frequency can we reasonably expect the number of Sixes to be?

De Moivre developed a simple formula that was widely useful for questions of this nature. One of his superb insights was to realize that the deviation of the actual number of Sixes from the average expected was best described by comparing it to the *square root* of the number of rolls.

It is hard to overplay the significance of this discovery. When you hear that an opinion poll has put support for a political party at 40%, it is often accompanied by a reminder that this is only an estimate, but that the true value is 'very likely' to be in some range like 38% to 42%. The width of such a range tells you about the precision of the initial figure of 40%, and if you want higher precision, you need a larger sample: this square root factor means that to *double* the precision, the sample needs to be *four times* as large! We have a law of diminishing returns with a vengeance – to do twice as well, we must spend four times as much.

De Moivre's approach can be illustrated by looking at how many Heads will occur in twenty throws of a fair coin. Taking all sequences of length 20 such as HHHTH... HTHT as equally likely, we can construct Figure 1, where the heights of the vertical bars show how many of the one million or so different sequences produce exactly 0, 1, 2,...,19, 20 Heads. The respective objective probabilities are then proportional to these heights. De Moivre showed that the best-fitting continuous smooth curve through the tops of these bars is very close to a particular form, now often called the *normal distribution*.

A curve of this nature arises for any large number of coin throws, and also when the chance of Heads differs from one half. All these curves bear a simple relation to each other, so de Moivre could produce a single numerical table for just one basic curve, and use it everywhere. A good estimate of the proportion of times that the overall frequency of Successes would be within certain limits could now easily be found – all that was needed was the chance of Success, and the number of times the experiment was to be conducted. You're going to roll a fair die 200 times and you want to know how likely it is that the number of Sixes will be between

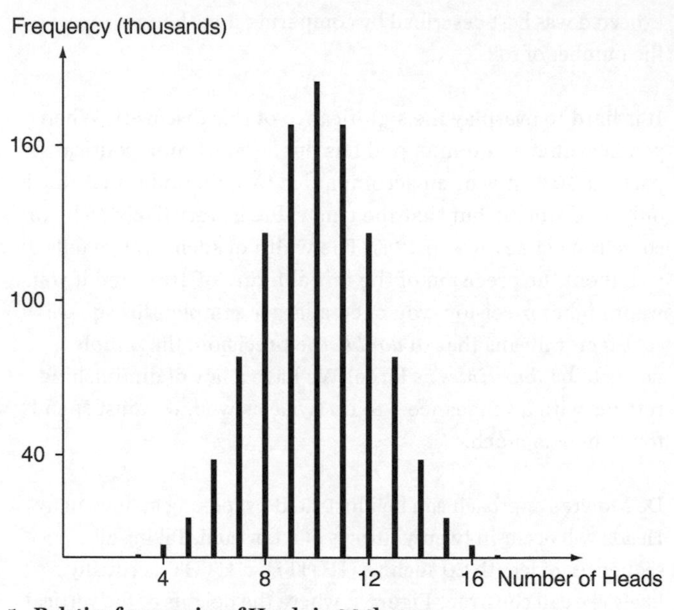

1. **Relative frequencies of Heads in 20 throws**

30 and 40? Or how likely is it that a fair coin will fall Heads more than 60 times in 100 tosses? No problem – de Moivre had the solution.

Suppose we know the ages of death for a group of men, all of whom reached at least their fiftieth birthday. De Moivre's work could answer the question: 'If a man aged 50 is more likely than not to die before reaching 70, how likely is it that the figures observed for that group would arise?' Useful though this was, it did not answer the key question posed by the nascent life insurance industry: 'How sure can we be that a 50-year-old man is more likely than not to die before he reaches the age of 70?'

Inverse probability

The ideas of Thomas Bayes, a Presbyterian minister who dabbled in mathematics, are far better appreciated now than in his lifetime. His *Essay towards solving a problem in the doctrine of chances*, published in 1764, three years after he died, gives the beginnings of a general approach to subjective probability, and a way of addressing the actuaries' problem about inferring probabilities from data. It also included an essential tool for working with probabilities, termed Bayes' Rule.

To illustrate the latter, suppose we throw a fair die twice. Given that the score on the first throw is three, it is easy to find the chance that the total score is eight, as this happens precisely when five is scored on the second throw. With hardly a pause, we give the answer 1/6. But turn the problem round, and ask: given that the total score is eight, what is the chance that the first throw yielded three? The answer is far less obvious, but can be found by applying Bayes' Rule. Under the standard model of dice throws, the chance turns out to be 1/5.

This notion of *inverse probability* is central to the way evidence should be considered in criminal trials. Suppose fingerprints found at a crime scene are identified as belonging to a known individual, Smith. The probability of finding this evidence, if Smith is innocent, is likely to be very low. But it is not 'How likely is this evidence, given that Smith is innocent?' that the Court passes judgment on: it is 'How likely is Smith to be innocent, given this evidence?' Bayes' Rule is the only sound way to obtain an answer. We will see in later chapters how this Rule helps in making sensible decisions.

The insights shown by Bayes were overlooked for many years, but he did identify the central problem: if the chance of Success in a series of Bernoulli trials, like dice throws, is unknown, but the respective numbers of trials and Successes are known, how likely is it that this unknown chance falls between specified limits? Laplace, a far superior mathematician, was able to carry out the computations that had defeated Bayes.

From tentative beginnings in 1774 to a synthesis in 1812, Laplace steadily improved his analysis, and gave explicit formulae to answer Bayes' question. For example, using data on the numbers of male and female births in Paris, he concluded that it was beyond doubt that the chance of a male birth exceeded that for a female – he put the probability this was false as about 10^{-42}!

Bayes in buried in the London cemetery of Bunhill Fields, near the Royal Statistical Society. The vault has been restored, and displays a tribute to Bayes paid for by statisticians worldwide.

The Central Limit Theorem

Write the list of outcomes of a collection of Bernoulli trials as a sequence of Successes and Failures, e.g. FFFSF FFSSF SFF... Now replace each S by the number one, and each F by zero, giving 00010 00110 100... This indicates a cunning way to think about the total number of Successes in these trials: it is just the *sum* of these numbers (agreed?). De Moivre had given a good approximation that described how this sum would vary, using his so-called normal curve.

A vast array of quantities we might want to consider do arise as a *sum* of randomly varying individual values. For example, a local authority responsible for rubbish disposal is interested mainly in the total amount over the town, and not in the separate random amounts from each household. When a gardener sows runner beans, it is the total yield, not that in each pod, that concerns him. A casino judges its financial success on its overall winnings, irrespective of the fates of individual gamblers. Being able to regard an item of interest as the sum of a large number of random bits is often fruitful.

Laplace extended de Moivre's work to cover cases like these. He established a *Central Limit Theorem*, which says that something that is the sum of a large number of random bits will, to a good approximation and in a wide range of circumstances, fit de Moivre's normal distribution. We don't need the details of how the individual components tend to vary, the way the *total amount* varies will closely follow this normal law.

To use this idea, we require just two numbers: first, the overall average amount, and second a simple way of expressing its variability. Given those two numbers, any probability can be found from de Moivre's tables.

Enter Carl Friedrich Gauss (1777–1855), bracketed with Newton and Archimedes at the top of the mathematical tree of genius. He was investigating how to deal with errors in the observations of the positions of the stars and planets. He suggested that on average the error was zero – observations were just as likely to be wrong a bit to the left as a bit to the right – *and* its size followed this same normal distribution. He took this path for its mathematical simplicity, but when Laplace saw Gauss's book, he linked it to his own work. He argued that because the *total* error in an observation arises as the agglomeration of many random factors, such an error *ought* to follow the normal law. Gauss's lame excuse of 'mathematical convenience' was replaced by Laplace's more persuasive 'mathematics indicates that…'.

The term 'normal', applied to this distribution, is unfortunate. It suggests that, in the first instance, we should expect any data we come across to follow its format, but this is far from the case. To avoid this implication, and to honour a great man, we will switch to the alternative term *Gaussian*. If you can persuade yourself that your item of interest can plausibly be regarded as the sum of a large number of smaller variable bits, having largely unrelated origins, this Central Limit Theorem says that the item can be expected to vary in a Gaussian manner.

Do observational errors really follow this law? According to Henri Poincaré, the last mathematician to feel comfortable across the whole existing mathematical spectrum, 'Everybody believes in it, because the mathematicians imagine it is a fact of observation, and observers that it is a theory of mathematics.'

Siméon Denis Poisson

Poisson is best known for a *distribution* – the way in which probabilities vary around an average – that carries his name. An example arose in the work of the physicist Ernest Rutherford and his colleagues, when they counted how many alpha particles were emitted from a radioactive source over intervals of length 7.5 seconds. This number varied from zero to a dozen or so, with an average just under four. Figure 2 illustrates two typical experiments, showing (in those cases) four/five emissions. Rutherford expected the emissions to occur at random.

Chop the 7.5 seconds into a huge number of really tiny intervals, so small that we can neglect the possibility of more than one emission in them. All but a few intervals will have zero emissions, the rest will have just one. Within each tiny interval, regard an emission as a Success, so that the total number of particles emitted is just the number of Successes – Bernoulli trials again.

In a really tiny interval, the chance of a Success is effectively proportional to its length, so as this length shrinks, we have an

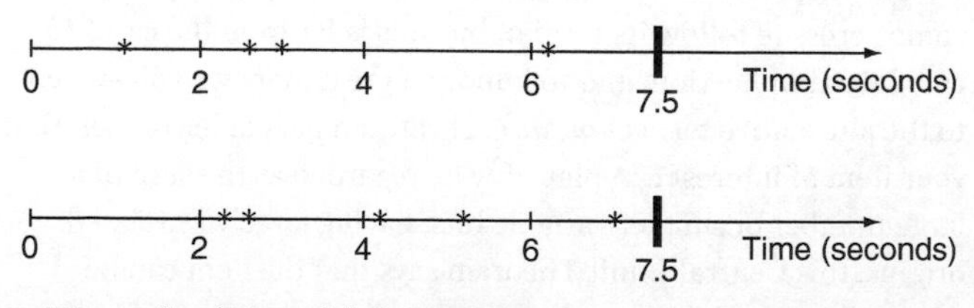

2. Times of emission of alpha particles

increasing number of intervals, each having a decreasing chance of Success. Poisson worked out the exact chances for 0, 1, 2,... emissions altogether as the lengths of the tiny intervals reduce down to zero.

This *Poisson distribution* arises frequently, at least as an excellent approximation, whenever the things we count are happening 'at random'. It was appropriate for Rutherford's data; it fits the numbers of flying bombs that landed on different parts of south London in World War II; it seems useful as a model for the number of misprints in each block of 1,000 words in a book. If you simultaneously deal out two randomly shuffled decks of cards, face up, on average you will have exactly one match between them; but the actual number of matches will vary very much like a Poisson distribution. A gruesome example of this distribution, foisted on generations of students, is of twenty years of data for the numbers of officers in the different Prussian Cavalry Corps who were kicked to death by their horses.

All of those examples conform to the same pattern: a large number of opportunities, each with a tiny chance of coming off. Whenever the phenomenon you are studying fits that template, this Poisson model is likely to be useful.

The Russian School

A mathematical theorem takes the format: if certain assumptions are true, then a desired conclusion follows. The main interest is in applying the desired conclusion, so it is most useful when the required assumptions are not very onerous. Sometimes the desired conclusion can be demonstrated only under very restrictive assumptions, or with great difficulty: later workers may find easier ways to use the same assumptions, or reach the same conclusion under less restrictive conditions. Best of all is when the conclusion can be shown true under very mild assumptions, and with a short and elegant argument. The work of Pafnuty Chebychev (1821-94) gives a fine example of this ideal.

Chebychev helped to show how a Law of Large Numbers applied in wider circumstances. The original Law related to Bernoulli trials, describing how well the proportion of Successes in a sequence of trials could estimate the chance of Success. If we want to estimate the average height of soldiers joining an army, or the cost of feeding a family for a week, it seems obvious that we can do so by taking a suitable sample from the relevant population. But how good will that estimate be? Chebychev's work gave a firm idea of the probability that the error would be small enough to make the estimate reliable.

Much of Statistics rests on the applications of these ideas.

Chebychev's best-known student is Andrey Markov, whose teaching inspired a further generation of talented Russians. Markov applied his ideas to poetry and literature. By replacing the vowels and consonants in Pushkin's *Eugene Onegin* by the respective letters v, c, he generated a sequence with just those two symbols. In the original Cyrillic alphabet, vowels formed about 43% of the text. After a vowel, another vowel occurred some 13% of the time, while after a consonant, vowels arose 66% of the time. To predict whether the next symbol would be v or c, he discovered that, given the current symbol, he could effectively ignore all its predecessors, so little help did they give.

This 'forgetting' property holds widely. Examples include: the successive values of a gambler's fortune; the daily weather (Wet or Dry) in Tel Aviv; the lengths of many queues, counted as each customer leaves; the genetic compositions of consecutive generations; the diffusion of gases between two linked containers. Whenever in a randomly varying sequence where we wish to predict future values, we can, knowing the present, ignore earlier values, the sequence is said to have the *Markov property*. The theory of such sequences is now well developed, and is the basis for many successful applications of probability ideas.

Markov, also active in politics, had a fine sense of mathematical history. In 1913 the Russian government organized celebrations to mark 300 years of Romanov rule, so Markov countered with events to commemorate the 200 years since Bernoulli's discovery of the first Law of Large Numbers.

I digress to mention the work, early in the 20th century, of the Frenchman Émile Borel. Recall the Law of Large Numbers for Bernoulli trials: that after many trials, it is overwhelmingly likely that the actual frequency of Successes is very close to the probability of Success. This still leaves open the possibility that, during an indefinite number of trials, the actual Success frequency occasionally ventures outside any given tolerance band around that Success probability. But Borel's work killed that notion stone dead. Given any such tolerance band, there will come a time (we can't be sure when, but it will happen) after which the actual frequency of Successes *stays inside* the band permanently. This is known as the Strong Law of Large Numbers.

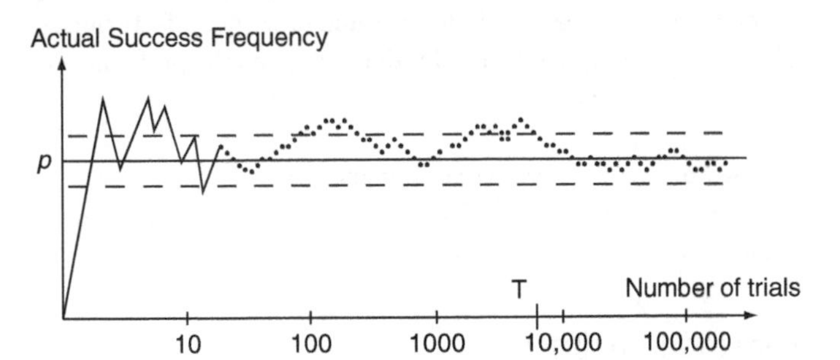

3. Illustration of the Strong Law of Large Numbers; *p* is the probability of Success, the dashed lines show a tolerance band. After trial T, the actual Success frequency stays permanently within the tolerance band

This Strong Law also extends to wider circumstances: we can sum up the message of the Laws of Large Numbers by the informal phrase:

> in the long run, averages rule.

In 1924, Alexander Khinchin published the wonderfully named *Law of the Iterated Logarithm*. Like the earlier work of Bernoulli and Laplace, when this was applied to a random quantity that arose as a sum, it could give even more precise information on how close that sum would be to its average value.

For some three hundred years, advances in the workings of probability came from a range of ad hoc methods. Then in 1933, the outstanding Russian scientist Andrey Kolmogorov used the recently developed ideas of *measure theory* to set the subject in a satisfactory logical framework. All the known theorems could be recast in Kolmogorov's setting, giving a precision that was a catalyst for the developments that followed.

Kolmogorov, along with Khinchin and their student Boris Gnedenko, also greatly extended Laplace's work on sums of random quantities. They were motivated by ways of increasing the reliability of machines used in textiles and other manufacturing industry, quality control on production lines, and the problems caused by congestion.

Kolmogorov was a superb researcher and teacher. When he died in 1987, the then Soviet president Mikhail Gorbachev rearranged his duties so as to be able to attend the funeral.

More modern times

War has frequently provoked scientific advances. The 1939–45 conflict boosted the development of operations research, with much of its success resting on sensible use of the ideas of probability. To maximize the probability that a supply ship would avoid being sunk by enemy submarines, a combination of data and calculation led to the conclusion that convoys were better than single ships, and large convoys better than smaller ones. When this conclusion was acted on, losses fell dramatically. The outline of the codebreaking work in Bletchley Park is now well known: however, the importance of the use of Bayes' Rule to identify the most promising path to discover the settings of the reels on the Enigma machines is often overlooked.

In 1950, William Feller published his magnificent introductory book on probability, with further editions in 1957 and 1968. This book is my nomination for the best non-fiction book ever written. Directly and indirectly, with its mixture of intuition and rigorous argument, it led to a spectacular growth of interest in the subject. A little later, Joe Doob used the term *martingale* (which originally meant the gambling 'system' of doubling the stake after each loss) for a collection of random quantities where, loosely speaking, their average value at some future time is the same as the current value. He developed the main properties of martingales and closely related ideas: this work was widely useful, as it turned out that many collections of random quantities of practical interest fell within the scope of this theoretical investigation. Later we will illustrate how probability concepts have been usefully applied across a range of fields.

Many academic journals specializing in probability have been launched, some have spawned offspring, none report that they are short of material well worth publishing. The capabilities of modern computers have transformed the environment for calculating probabilities: their speed and storage capacity have greatly increased the range of soluble problems. Earlier, most work was when probabilities were influenced by just one factor, say time or distance, and exact calculation by humans was often possible; now, complex problems where probabilities change with time, three dimensions of space, and other influences, have been successfully attacked.

Even so, it may well be that the largest influence of computers on the development of probability is through ease of communication. The language $\TeX$ has become the standard framework in which mathematics and much of science is written up. Research workers post their thoughts and ideas on the internet, scholarly articles can easily be accessed from home or office on the World Wide Web.

Chapter 4
Chance experiments

Think of any experiment with chance outcomes – buying a lottery ticket, betting on a horse race, going on a blind date, undergoing some medical treatment. We use the word *distribution* to specify all the possible outcomes, along with their associated probabilities. (We slipped in that word when writing about Poisson's analysis of how many rare events will happen, given a large number of opportunities.)

The 'distribution' is central to analysing the range of consequences from a chance experiment. Plainly, we need to be clear about the full extent of the possible outcomes. To give sensible values for their probabilities, we must spell out our assumptions, and hope that they are appropriate for the experiment we seek to investigate.

Discrete distributions

First, we look at circumstances where the possible outcomes can be written as a list, each outcome having its own probability. The phrase *discrete distribution* applies here.

The most straightforward case is when we can count the number of outcomes, and agree that they should all be taken as equally likely. The term *uniform distribution* is used, as the total probability is spread uniformly over the outcomes. Many experiments are expected to fit this bill – roulette, dice, hands of cards, selecting the winning numbers in a lottery, etc. Accurate counting generates the appropriate answer.

Recall the term 'Bernoulli trials' to mean a sequence of independent experiments with a constant probability of Success each time. With a fixed number of Bernoulli trials, there is a simple formula, called the *binomial distribution*, that gives the respective probabilities of exactly 0, 1, 2,... Successes. This formula depends only on the number of trials, and the Success probability. As you run through the outcomes in order, their probabilities initially increase up to a maximum value, then fall away towards zero. (Poisson distributions also follow this pattern.)

We expect a binomial distribution for the number of Sixes among twenty throws of a die; or the number of correct answers when a student guesses randomly among five choices at each of thirty questions on a multiple choice exam. But we do *not* expect it when asking how many Clubs a bridge player has among his thirteen cards: although each separate card has probability one quarter of being a Club, successive cards are not independent, as the chance of a Club on the next card is affected by all previous outcomes.

Always read the small print. A binomial distribution requires *three* conditions: a fixed number of trials, each independent of the rest, and with a constant chance of Success.

In a sequence of Bernoulli trials, what is the chance it takes exactly five goes to achieve the first Success? The only way this happens is to begin with four Fails, then have a Success; and since all trials are independent, the answer comes from multiplying the respective probabilities of these outcomes together, giving a pleasingly simple expression, the so-called *geometric distribution*.

The probabilities of taking exactly 1, 2, 3, . . . trials for the first Success decrease steadily. Each time, the next probability comes from multiplying the present value by the chance of one more Fail, some fixed value less than unity. Thus, whatever the chance of Success, the single most likely number of trials to achieve the first Success is always unity!

Make a leap of faith, and suppose that, in cricket, successive balls form Bernoulli trials. A bowler, who interprets 'Success' as meaning that he takes a wicket, can think optimistically: when he comes in to bowl, the single most likely time he will take his next wicket is with the next delivery. Conversely, a batsman who takes a similar view must fatalistically accept that the most likely duration of his innings is that he faces just one ball. (Even for the best batsmen, records confirm that their single most likely total score is usually zero!)

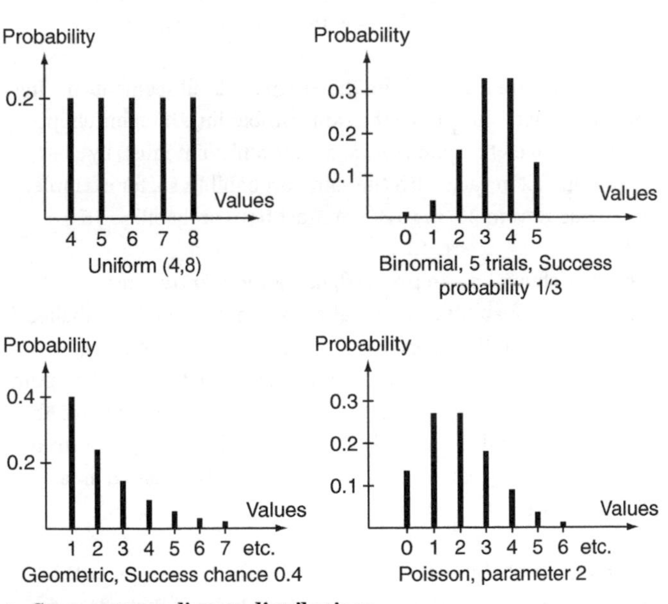

4. Some common discrete distributions

Figure 4 illustrates some of the common discrete distributions. For each possible value, the height of the vertical bar gives its probability, and the sum of all the heights is always, of course, unity.

Continuous distributions

How might we extend the classical ideas of probability to deal with the experiment of choosing a random point on a stick of length 80cm? Here there is a *continuum* of possible outcomes, not just a list.

'At random' means that all individual points have the same probability. But if that common value were to exceed zero, then, by taking sufficiently many points, their total probability would exceed unity, which is impossible. Each separate point must have probability zero, and we can no longer use pictures like Figure 4. Rather than associate probabilities with individual points, we need to associate probabilities with segments, or intervals.

To give equal treatment along the 80cm stick, all segments having the same length must have the same probability. Imagine chopping the stick into eight equal pieces: a 'random' point must, by definition, fall in each with the same probability, so, for example, the segment from 20cm to 30cm must have probability 1/8.

Figure 5a shows how to proceed, using the mantra 'Area represents probability'. The height of the horizontal line labelled h is chosen so that the shaded area beneath that line is unity, representing the fact that it is 100% certain that the random point falls *somewhere* along the interval from 0 to 80. Then Figure 5b shows how to find the probability of falling in the segment from 32cm to 52cm, by calculating the corresponding shaded area. Plainly, this is 1/4.

To find the probability that a randomly selected point is within 10cm of either end of the stick, or within 10cm of the centre, we could use Figure 5c, and appeal to the Addition Law. The required probability is the sum of the three shaded areas, namely one half.

Figure 6 illustrates a similar path for other situations where the outcome takes continuous values, such as the time until the next accident on a particular stretch of motorway. We will argue below that the general shape of the curve shown is reasonable in this situation, but the main point is that the scale is chosen so that the total area above the line marked 'Time', but below the curve beginning at the point E, is unity, as it is 100% certain that the time to wait takes some non-negative value.

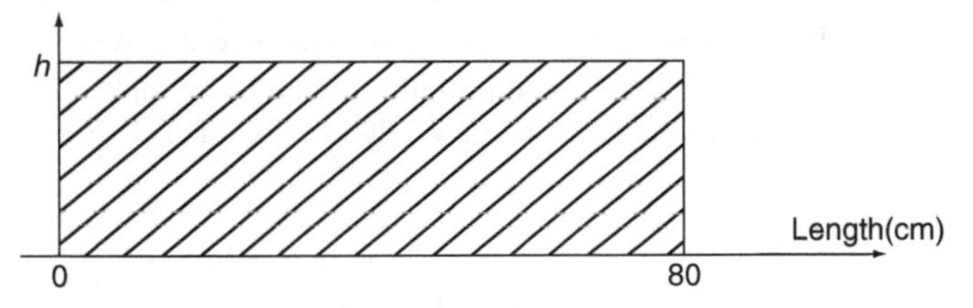

5a. The shaded area is unity

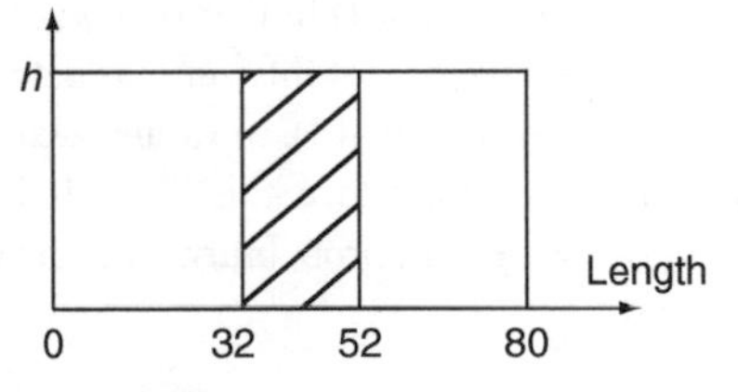

5b. The probability of falling between 32 and 52 is 1/4

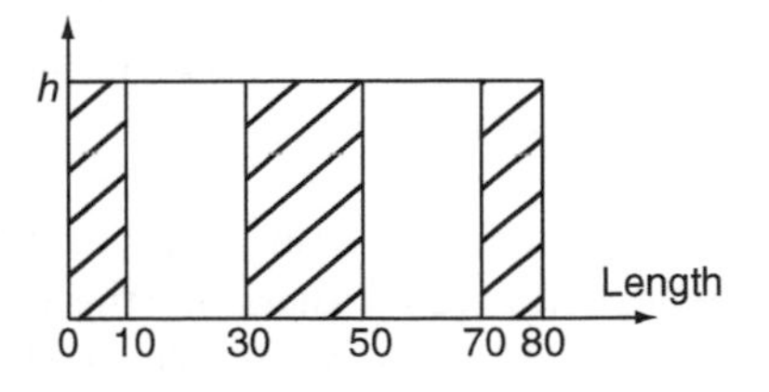

5c. See text

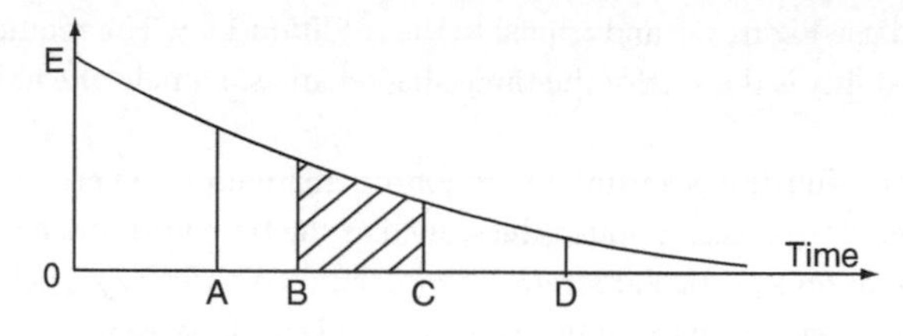

6. A continuous distribution

The probability that the time is at least B, but no more than C, is the size of the area shaded. In a similar fashion, we can find the probability that the time to wait falls in any given interval, and then, using the Addition Law as above, the chance it falls in more complex regions.

A curve that generates probabilities in this manner is called a *probability density*. Now area is calculated as 'length times breadth', and the breadth of any line is zero. Hence the 'area' of either of the vertical lines at A or D in Figure 6 is zero, so both those individual points have probability zero, as before. But the density curve is higher at A than at D, so values *near* A are more likely than values near D. At a glance, the Figure indicates the regions of relatively low or high probability. The term *continuous distribution* is used.

In all such experiments, since individual points have probability zero, we can be a little slipshod: whether an interval includes both endpoints, just one of them, or neither, the probability the outcome falls in it is the same.

To qualify as a probability density, a curve must have two properties: it cannot take negative values, and the total area underneath it must be unity. This ensures that all calculations of probabilities lead to sensible values.

Many probability density functions arise often enough for them to be given names. For the experiment of selecting a random point within a given interval, the density function will be completely flat over that interval, as in Figure 5: plainly, all segments of the same length do indeed have the same probability. Again, the term *uniform distribution* is used.

Suppose we are interested in the time to wait for some special event. For example, ^{210}Pb is an unstable isotope of Lead, and the claim 'Its half-life is 22 years' appears in physics textbooks. The meaning is that, whenever we take a lump of this substance, only half of it is unchanged after 22 years, the rest having decayed into other substances through radioactive emission.

This lump consists of a gigantic number of atoms, all acting independently. Focus on one atom: at some random time, it decays by emitting a particle. We do not know when this will be, but since half the atoms in the lump decay in 22 years, the chance that *this particular atom* decays within that time period is 50%. Suppose it has not decayed after five years: at that time, it is just one atom in the residual lump of ^{210}Pb, so the chance it decays within a further 22 years is again 50%. And if it has not decayed in the next three years, the same applies, and so on.

It turns out that the only way this can happen is when the random time until a given atom decays has what is known as an *exponential distribution*, whose density has the general shape shown in Figure 6, the height of the curve falling at a constant rate. A similar background applies to road accidents: if none has occurred in the past week, that seems unlikely to affect the chances of an accident in the future, so we expect the time to wait for a road accident also to follow an exponential distribution.

This distribution is intimately tied up with the Poisson distribution. Whenever things are occurring essentially at random – flashes of lightning in a storm, spontaneous mutations

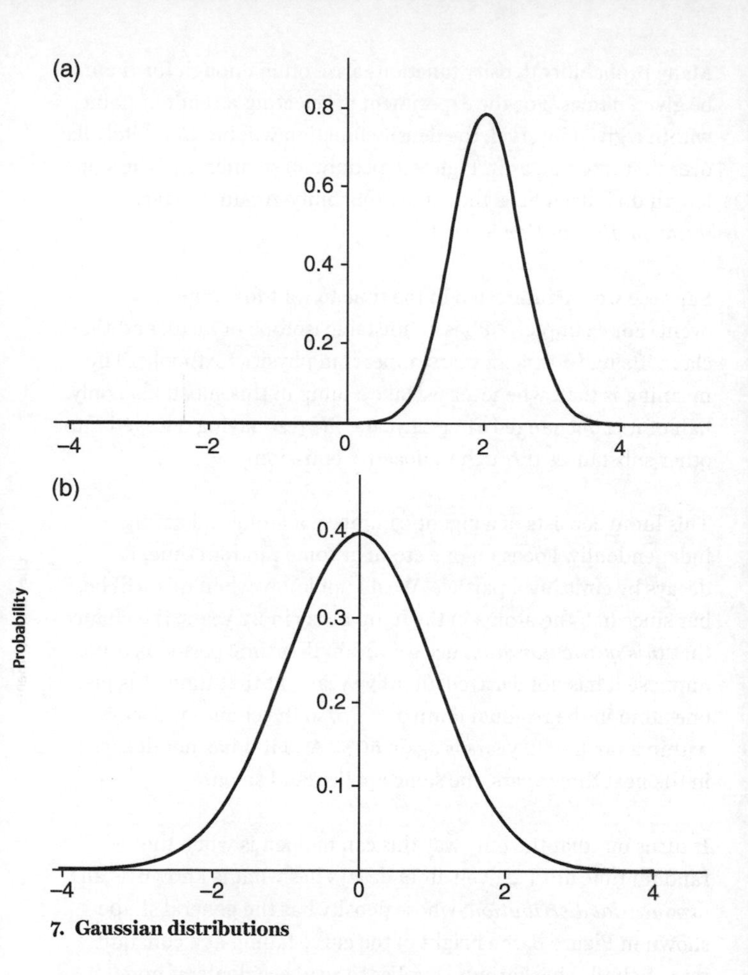

7. Gaussian distributions

in reproduction, the arrivals of some customer at the Post Office – the *number* of such events in a fixed time period tends to follow some Poisson distribution, while the *time to wait* between events has this exponential format.

The most important continuous distribution is the one we have already named the *Gaussian distribution*. As Figure 7 illustrates,

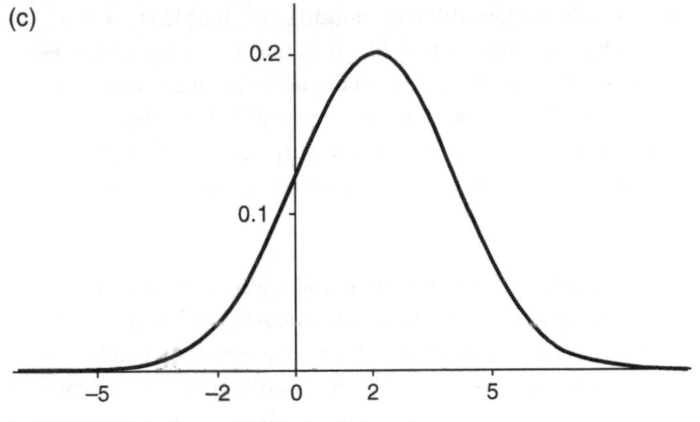

7. Continued

members of this family are symmetrical around a single peak and fall away rapidly towards zero, while never actually attaining that value. Two numbers tell us to which member of this family any example belongs: one number picks out the peak, the other describes the spread – small values of the spread lead to tall and narrow graphs like Figure 7a, larger values give short, fat graphs like 7c. Any probability for a member of this family can be found by using these two numbers to relate it to Figure 7b, which has its peak at zero and its measure of spread standardized at unity. Suitable tables have been widely available since de Moivre first produced them.

Resolving an issue

You may have noticed a problem. Provided the set of supposed outcomes is finite, or an unending list like {1, 2, 3, . . . }, then even if some members of this list turn out to have probability zero, any event whose probability is zero will never happen. However, with a *continuous distribution*, although each separate point has probability zero, one of them *will* occur when the experiment is performed! We can no longer take 'will not happen' and 'probability zero' as meaning exactly the same.

To reconcile matters, think of choosing one marble at random from a box holding a million identical marbles. We would be very surprised if we correctly guessed the outcome in advance, as the chance of doing so is only one in a million. But whichever marble gets chosen, we do not then express surprise, even though an outcome, whose probability was as small as one in a million, has occurred.

Make the box bigger – a billion marbles, a trillion – and the corresponding chance of the actual outcome can be made as close to zero as we like – but it did happen. Choosing one point at random on a continuous line is not very different from this: for any point, its chance is zero, but one of them will occur.

We will show below that, in a repeatable experiment where the chance of guessing the outcome is one in six, we can expect to perform that experiment six times in order to be right just once. Similarly, if the chance is one in a million, we expect to take a million repetitions to guess right once. Reduce the chance of occurrence by a factor of another million, and the time we expect to wait for a correct guess gets multiplied by a million. Outcomes with really tiny probabilities do occur, but more and more rarely.

If the probability falls all the way down to zero, we can expect to wait longer than any finite time – it just won't happen! It is rational to act as though any event of probability zero, that is *named in advance*, will not occur.

Mean values

Knowing the distribution of the outcomes from a chance experiment, we can calculate any probability we like. But sometimes, all this detail gets in the way – we can't see the wood for the trees: so we want to pick out the main features of the distribution.

To illustrate, suppose the only outcomes possible are 2, 3, and 7, with respective probabilities 60%, 10%, and 30%. We expect that, over a hundred repetitions of this experiment, the value 2 will occur some sixty times, 3 about ten times, and 7 the remaining thirty times. The total of all these values is 120 + 30 + 210 = 360, so the average over all the one hundred outcomes is 360/100 = 3.6. This answer is just the *weighted* sum of the values 2, 3, and 7, the weights being their probabilities.

Whatever distribution we have, similar calculations lead to the average outcome over a large number of repetitions. 'Average' is a loose word, we prefer the term *mean* for the result of this calculation. There may be short cuts: if the values are *uniformly* distributed over some range, the mean is just midway between the two extremes; the mean number of Successes in a sequence of Bernoulli trials comes from multiplying the number of trials by the chance of Success.

When rolling a fair die, the chance of getting a Four is 1/6. So among 600 throws, we should see around 100 Fours: simple arithmetic then says that the mean *gap* between successive appearances of a Four is 6. It is plainly no coincidence that a chance of size 1/6 leads to a mean gap of 6. The length of any gap is just the time to wait for the next Success, so we have the pleasing result that, during a sequence of Bernoulli trials,

> the mean time to wait for a Success is the reciprocal of the probability of Success.

With continuous distributions, the idea is the same, but the weighted sum is found by using the mathematical technique known as *integration*. For Gaussian distributions, the mean is where the peak occurs. Exponential distributions arise as the time to wait for a random event, which occurs at some characteristic overall frequency: it should be no surprise that the mean time to wait is just the reciprocal of that frequency.

The terms 'expectation' and 'expected value' are also used instead of 'mean' and 'mean value'. Tossing a fair coin a dozen times, the 'expected' number of Heads is six, and the 'expected' score when throwing an ordinary fair die is 3.5. Of course, just because the expected number of Tails on a single toss is 0.5, you don't actually *expect* to get half a Tail! The English language has many quirks.

Means are very friendly animals: the mean of a sum is always the sum of the means, whether or not the different components arise independently. The Law of Large Numbers says that, in the long run, means dominate: if you spend £1 on a Lottery ticket, where half that sum goes into the prize fund, then, however the prize distribution is structured, your *mean* return is 50p and, in the (very) long run, that is what you will get.

Variability

It is also useful to have a succinct way of describing the variability of a distribution. We could calculate the difference between each value and the mean, and then find the (properly weighted) average value of these differences. But, as any trial calculation will show, this path is fruitless: the negative differences inevitably exactly cancel the positive ones, always giving a final answer of zero.

But whether a difference is positive or negative, we will get a positive number when we square it. So we could use the weighted average of these squared values to assess the variability. This quantity is called the *variance*. If the distribution is concentrated near the mean, the variance will be small; it will be larger when there is a reasonable chance of getting values well away from the mean.

When considering income distributions, with data in dollars, the squared data are in 'square dollars', whatever that might mean. Taking the square root of the variance returns us to the original measurement units, giving what is called the *standard deviation*.

The mean and standard deviation together often give a swift and helpful way of picking out main features of a probability distribution. And in the Gaussian case, these two numbers suffice to find any probability at all! As useful touchstones, the outcome when the distribution is Gaussian will be within one standard deviation of the mean about 68% of the time, within two standard deviations over 95% of the time, and only one time in 400 will it be more than three standard deviations away.

These figures are the basis for the guidelines offered in Chapter 1 about how close an agreement we can reasonably expect between Success probability, and the actual frequency of Success: the key is the Central Limit Theorem, which says that quantities that arise as the sum of a large number of random components are expected to follow a distribution close to the Gaussian.

In Figure 7, showing three Gaussian density functions, the means of the graphs are at 2, 0, and 2, while their respective standard deviations are 1/2, 1, and 2.

But be warned: although the mean of a sum is always the sum of the means, the same is not generally true of either the variance or the standard deviation. If the components of the sum happen to be independent – say a casino's profits over seven separate days in Las Vegas – then the variance of the sum will indeed be the sum of the individual variances, but otherwise it could be higher or lower. Adding standard deviations together seldom leads anywhere sensible.

Extreme-value distributions

In several applications of probability, interest centres on the largest or the smallest of a large number of random quantities. For example, the strength of a thread or a cable rests on the properties of the weakest fibre; flood defences take account of the maximum surge that might be expected over the next hundred years; the subject of *survival analysis* examines what fraction of a population remains after a given time. Extreme events may occur rarely, but when they happen, the consequences can be important.

The simplest plausible model assumes there are independent random quantities, each following a particular distribution; for example, the claims made on an insurance company in each separate year. The company is interested in how big is the largest total claim it can expect to receive over the next fifty years. There is a useful mathematical result that goes a long way to answering this question: however the claims vary over a single year, there are only *three* possible types of answer for the maximum claim over a large number of years. They are known as the extreme-value distributions, with the specific names of the Fréchet, the Gumbel, and the Weibull distributions. There is a sound mathematical principle that if there is some theorem about maxima, there is a corresponding result about minima. So if the item of interest is some minimum value, the same conclusion pertains.

To be able to limit the possibilities to these three families of distributions is very helpful. By estimating the mean and variance of an extreme value, and selecting whichever of them seems to fit the data best, sensible estimates of other probabilities – the chances of really extreme and devastating events – can be found.

Chapter 5
Making sense of probabilities

I will suggest how probability ideas can help in making decisions in the face of uncertainty, and also describe circumstances where misunderstandings can arise.

Odds?

Recall that probabilities can be expressed in terms of odds, and vice versa: a probability of 1/5 is the same as odds of 4 to 1 against. Unfortunately, the term 'odds' has also been usurped by the gambling community to mean something quite different – the amount the bookies will pay if your selected horse wins. So when you read that Sea The Stars won the 2009 Derby at odds of 11 to 4, that simply means that for each £4 staked on the horse, the profit, because it won, is £11. The figures '11 to 4' have no automatic relationship with the probability of winning. They depend on the bookies' subjective assessments of the horse's chances, and on how much money gamblers have staked. The term 'payout price' is a more accurate use of language for these figures like 11 to 4, but, regrettably, we have to accept the common usage of 'odds' in this gambling context.

A payout price is termed *fair* if it gives no monetary advantage to either party, i.e. the mean value of the gamble is zero. The fair payout price for correctly picking the suit of a card selected from a well-shuffled deck is 3 to 1, as those are the exact odds against a correct guess.

Commercial gambles are not fair, in this sense, as they could never operate without a house advantage. For roulette in a UK casino, when all 37 outcomes are equally likely, the payout price for betting on a single number is only 35 to 1, not the 36 to 1 that would be fair. So the *mean* return on a bet of £37 is £36, giving a house advantage – the percentage of any bet it expects to win – of 1/37, about 2.7%.

This advantage is the same for most of the available bets in roulette: whether you are betting on pairs of numbers, triples, groups of four, or six, or twelve, for every £37 you stake, your mean return is always £36. But in Las Vegas, the standard house advantage is bigger, because of an additional slot, double zero, giving 38 outcomes – with the same payout prices as in the UK. The mean return on $38 is generally $36, a house advantage of 2/38, or 5.3%.

A different way to bet on horse races is through a Tote or pari-mutuel system. Here, the money bet on all the horses is pooled, and a fixed proportion – 80% or so is common – is shared among those who backed the winner, in a size proportional to their stakes. The Tote advantage is then 20%, whichever horse wins.

The size of the bookies' advantage in a horse race depends on which horse happens to win. Although bookies may make a loss or profit on any single race, recent data tell a sobering story: at a payout price of 6 to 4 on, punters should expect to lose about 10% of their stake; at a price of 5 to 1, expect to lose about 13%; at 10 to 1 the mean loss figure is over 23%, and if you speculate on horses priced at 50 to 1, expect to lose about two-thirds of your money.

This phenomenon is known as the *favourite-longshot bias*. Punters lose their money *more slowly* from bets on the more favoured horses than if they are attracted by large payout prices. Bookmakers were delighted when Mon Mome won the Grand National in 2009 at a price of 100 to 1.

Absolute risk, or relative risk?

Suppose that, among a particular group of people, the chance of developing colorectal cancer over the next five years is quoted as one in a thousand. We expect about ten among 10,000 people to develop the cancer. A new drug would reduce the chance to one in two thousand: then only about five among 10,000 would succumb if the new drug is used. The drug company could headline its press release 'Risk of cancer cut by 50%'. And that is accurate: for each person, the risk would be halved.

This approach describes a *reduction in relative risk*, and is often criticized as putting too favourable an interpretation on the data. For, suppose the initial risk had been one in ten million: cutting it by 50% leads to a new risk of one in twenty million, but in either instance, the risk is so small that among 10,000 people, we would expect pretty much the same number of cases – zero. Despite the risk being halved, the drug would hardly ever make a difference.

But suppose members of this group had a much higher chance, say 40%, of developing the cancer. A drug reducing the chance to 20% would qualify for the same headline, and would correctly be hailed as a major breakthrough, as among 10,000 people, fully 2,000 fewer would develop the cancer.

Rather than focus on the relative risk, it is usually more meaningful to look at the change in absolute risk. In the first case above, the absolute risk changes from 0.1% to 0.05%, so the drop is 0.05%; in the second case, the drop is a minuscule 0.000005%, while with the final figures, the drop is an impressive 20%.

A sensible way to proceed is to state the mean number of patients who should take the drug in order to prevent one case of the disease – the *Number Needed to Treat*, or NNT. The fewer the better, and this number is just the reciprocal of the change in absolute risk. In the examples above, the respective NNTs are two thousand, twenty million, and five.

Treating twenty million people to prevent one case of a disease is hard to justify. The NNT, along with knowledge of the treatment costs and the severity of the impact of the disease, allows us to make sensible decisions about allocating health care resources.

Combining tiny probabilities

How likely is it that at least one among an enormous number of events, each having a tiny probability, will occur? This can be the pertinent question when considering the probability of a catastrophe. Complex systems or machinery may fail if any one of a myriad of components fails; will two aircraft collide, or might a nuclear power station suffer a meltdown? The so-called *Borel-Cantelli Lemmas* give some pointers. These mathematical results show that, in many circumstances, the key quantity is the *sum* of all those tiny probabilities: if it grows without bound, catastrophe is certain.

One consequence is that we can never be satisfied with current safety standards. It is essential to continue to make improvements. For, no matter how high our standards, there is some non-zero probability of failure during any given month: and however small this value, if it remains unchanged (or even if it decreases too slowly) the sum over many months will grow indefinitely large, and disaster will occur sometime.

A sound programme of continual improvement does not guarantee that disaster will be averted: but to be ever satisfied with the status quo is to invite doom.

Some misunderstandings

(a) When a doctor tells a patient that there is a 30% chance that a particular medication will have unpleasant side effects, he means that he expects about 30% of patients on this drug to suffer. However, the patient may believe that these effects will arise on about 30% of the occasions on which she takes the drug. The doctor is thinking about all the patients he sees, the patient about all the times she takes pills – their *reference classes* are different.

(b) How does the public interpret the claim 'There is a 30% chance of rain in Chicago tomorrow' from a TV weather forecaster? The forecasters expect their audience to make a frequency interpretation, i.e. that, in the long run, rain would fall next day on 30% of the occasions when the weather conditions were similar to those now seen.

But when questioned, even among those viewers who were happy with the phrase 'a 30% chance', there was a spread of beliefs. Some felt that they were being told that rain would fall over 30% of the city's area; others that it would rain in Chicago for 30% of the day; and some believed that 30% of meteorologists expected it to rain! A few thought that it would definitely rain, with the 30% figure indicating the rain's intensity. There were many mismatches between the event the forecasters were referring to, and the event viewers were thinking about.

(c) If as few as twenty-three people chosen at random gather together, it is more likely than not that two of them share a birthday. When people meet this fact for the first time, they are normally surprised, but usually become convinced when proper counting shows this claim to be true. However, a minority remain unconvinced, because they mistakenly think they have been told that, if they and twenty-two others gather together, it is more likely than not that one of the others shares *their* birthday. Listen carefully!

(d) Suppose that a coin, accepted as fair, shows Tails on nine consecutive tosses. Some will claim that the next toss is almost certain to be Heads, perhaps by invoking some 'Law of Averages' that requires Heads to eat into this excess of Tails immediately. No such Law exists. The Law of Large Numbers does imply equal proportions of Heads and Tails, but only *in the long run*: any sequence of nine Tails is diluted by the thousands of tosses before and after.

Alternatively, some will (correctly) note that the chance of ten consecutive Tails is less than one in a thousand; so if they see nine consecutive Tails, they may then 'deduce' that Tails next time is highly unlikely. But that is false logic: if ever we have nine Tails in a row, a tenth Tail will happen half the time. This confusion of the absolute probability of an event, with its conditional probability, given a certain background, famously arose in 1996: jockey Frankie Dettori rode the winners of the first six races at Ascot, and since no-one had ever won all seven races in a day, it 'must' be virtually impossible for Dettori to win the last race. But he did. Very few people will ride the first six winners in a seven race meeting, but when someone does so, he might well also win the last.

Ask yourself: am I assessing the absolute chance of twenty things happening, or just the conditional chance of the twentieth, given that the first nineteen have happened?

(e) Newspapers are often produced under great time pressure, so it is not surprising that some articles contain nonsense. But here are three reports that should have been spiked.

When all six eggs in a box bought in a supermarket were double-yolked, it was claimed that a truly astonishing event had happened. Only one egg in a thousand has this property, so the chance of getting a boxful is this tiny fraction, multiplied together six times. The resulting number is so small that if you opened one box every second, you would expect to take over thirty billion years to come across a box containing only double-yolked eggs!

But that calculation makes sense only if all the eggs in a box are chosen independently from a vast collection of eggs, one in a thousand of which is double-yolked. This doesn't happen. Eggs are sorted by size before being boxed. Some boxes are even labelled as containing only double-yolked eggs....

Allegations about the private life of an England soccer captain surfaced. A reporter offered figures for the 'likelihood' of each of four possible actions by the team manager:

(a) expel him from the playing squad – 1/10
(b) retain in the squad, but invite him to resign – 3/10
(c) retain in the squad, but remove the captaincy – 6/10
(d) take no action – 8/10

Any single one of these four estimated probabilities is plausible. But since they relate to mutually exclusive outcomes, their sum must not exceed unity: however, these 'probabilities' add up to 1.8.

Thirdly, it was reported that, among people who had won at least £50,000 on the National Lottery, the names turning up most often included John, David, Michael, Margaret, Susan, and Patricia. So far so good: but it was absurd to claim that *therefore* you should try to include people with those names in your Lottery syndicate!

Describing ignorance

Drawing one card from a well-shuffled pack, I expect Black and Red to be equally likely, and I confidently attach the figure of '1/2' to the chance of Red. In a slightly convoluted way, I could say that the *distribution* I attach to the chance of getting Red allocates 100% to the quantity '1/2' – my confidence is shown by the choice of 100%. If I am sure that a probability is some definite figure, I attach probability 100% to that figure.

But frequently, I cannot select a single figure: my best estimate of the chance my train will miss its connection may be 3/4, but 2/3 and 4/5 could be nearly as plausible, and even values close to the extremes of zero and unity are not completely ruled out. I can use a continuous distribution over the range from zero to unity to describe my feelings about this unknown probability.

Total ignorance of a probability – very rare circumstances – would be described by using the continuous uniform distribution of Figure 5. More often, there is some intermediate figure which is your best single guess at the probability, and your instinct about both higher and lower values is that their chances fall away towards zero. Figure 8 shows a collection of graphs from what is called the *beta* family of distributions that have this property.

Figure 8a indicates that we are pretty ignorant of the value of the probability: we attach highest belief to values near 1/2, but values as small as 1/5 or as high as 4/5 are still quite possible; with Figure 8b, we are much more confident that the value is close to 1/2, while still not ruling out the extremes. With 8c, our highest belief is for values close to 1/3, while with 8d the values are very strongly congregated near 2/3, with very little expectation that the value is below 1/2.

How far will my car travel on 10 litres of petrol? When warmed up, and driven at a steady speed, I expect to get about 90 miles, but if I make short journeys over a couple of weeks, it will be more like 60 miles. In either case, the distance will have some uncertainty, expressed via some continuous distribution. To get a handle on what kind of distribution, imagine splitting the petrol into small cups of size 10cc. There will be 1,000 such cups, and the total distance covered will be the sum of the distances achieved using these 1,000 components. Recall the Central Limit Theorem, which says that the *sum* of a large number of random quantities will tend to follow a Gaussian distribution.

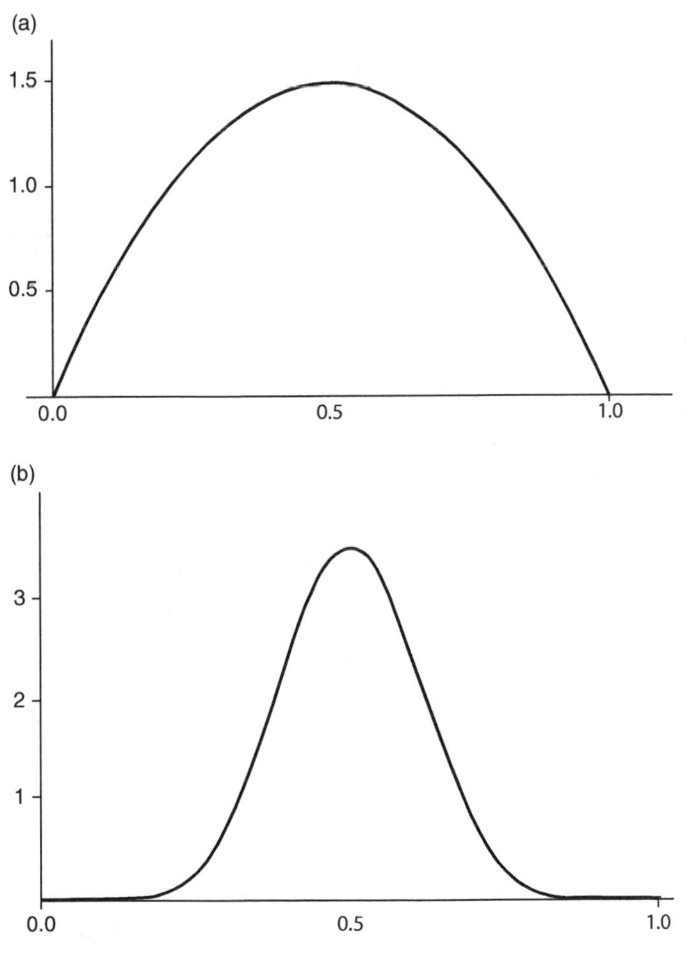

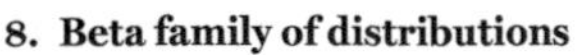
8. Beta family of distributions

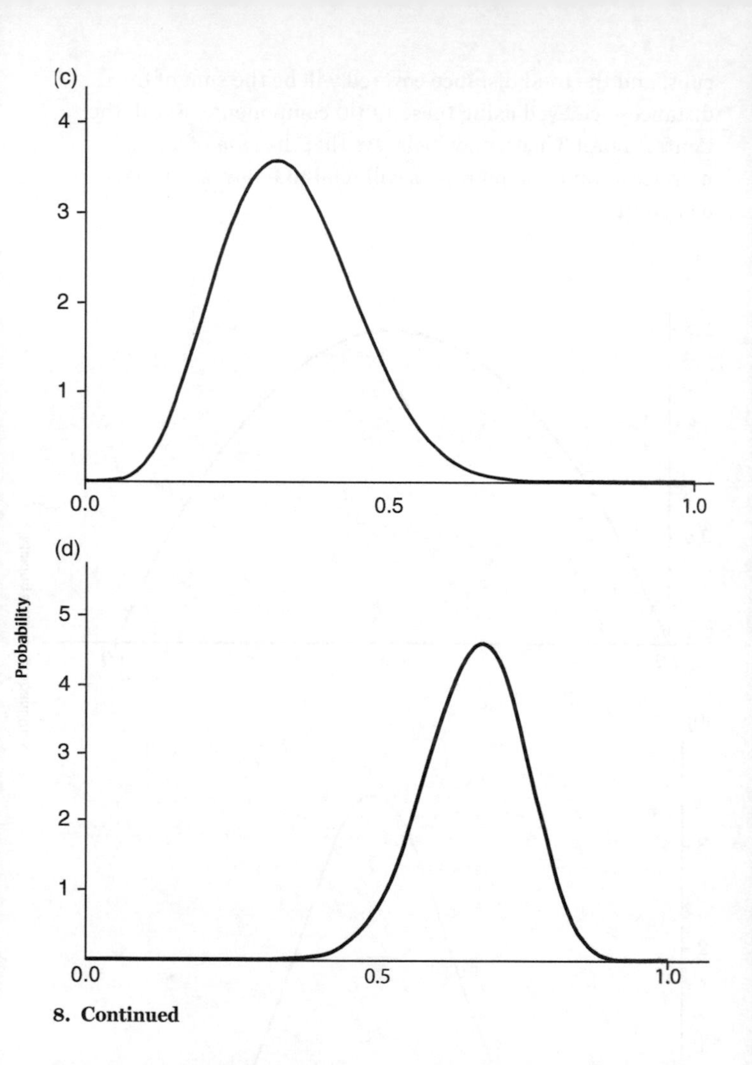

8. Continued

For steady motorway travel, I would select a Gaussian distribution centred on 90, sharply peaked to show low variance. For sporadic travel within town, I would also use a Gaussian distribution, but centred on 60 and with a wider spread to demonstrate the greater uncertainty.

Utility

Your fairy godmother makes you a once only offer. Either she will give you £1, or she will toss a coin and if you call Heads or Tails correctly, she will give you £10, otherwise you get nothing. What choice would you make?

The alternatives are £1 for certain, or a fifty-fifty gamble with a mean payout of £5. Nearly everybody prefers the latter. But scale the money up by a factor of one million: overwhelmingly, preferences change. To have one million pounds for certain is far more attractive than being equally likely to have zero, or ten million pounds. The concept of *utility* lies behind this difference.

For small sums of money, having twice as much usually *is* worth twice as much, but if one million pounds would generate a certain level of pleasure for you and your family, double that amount would not lead to double the pleasure. In whatever units you choose to measure the 'worth' of a monetary amount, the usual shape of the relationship will follow that shown in Figure 9: the graph always rises, initially like a straight line, but then steadily more slowly.

'Utility' explains why a householder and an insurance company can agree that a sum of £250 is a reasonable annual premium to

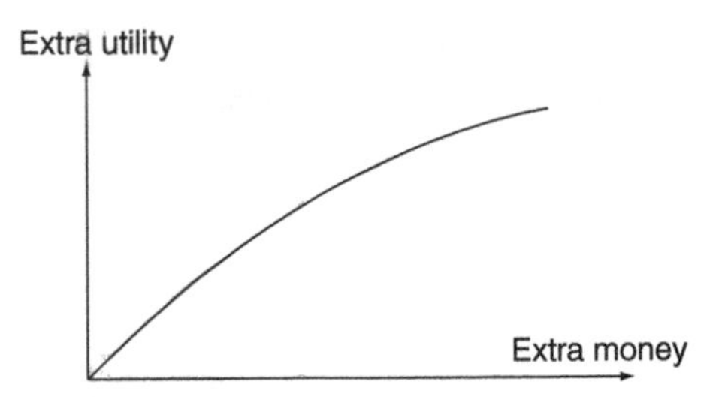

9. The general shape of a utility curve

insure a house worth £250,000 against perils such as fire, subsidence, or flooding. The substantial reserves of the insurance company mean that, on any individual house, it can act as though the utility is identical to the relatively small sums involved. So long as the chance that it will pay out in any year is less than one in a thousand, its expected value for this transaction is positive, it will make a profit. On the other hand, the uninsured householder would face the enormous negative utility of finding £250,000 if she lost her house to one of those perils, so voluntarily surrendering just £250 to remove that possibility is a good deal to her too.

Taking out insurance against breakdown of televisions, microwave ovens, and so on is almost always a bad idea. The sums are much smaller, utility and money are essentially the same, and the company will charge a premium large enough for them to expect a profit. Instead of buying this expensive insurance, build up a repair fund by placing the hypothetical premium into your bank account. Very few people will regret that move.

If you can successfully construct your own utility function, you can use it to help make a choice between the available actions under conditions of uncertainty. For each action, calculate the *expected utility* of the outcomes, i.e. weight the utilities by their respective chances. Then select that action for which this expected utility is as large as possible.

That is the probabilist's universal recipe for making the best of the available choices, whatever the circumstances.

Chapter 6
Games people play

Many recreational games combine skill and chance. Skill you can work on, chance is, well, a matter of luck. For all the 'games' discussed here, it is easy to persuade yourself that there is some finite list of outcomes, all equally likely. Thus in this chapter, unless explicitly stated otherwise, we will use the classical approach to finding probabilities: count the number of possible outcomes, and the probability of any event is taken as the proportion of those outcomes where the event happens.

My aim is to show how the ideas of probability can help a player make good decisions under conditions of uncertainty. An understanding of probability can also add to the enjoyment or entertainment of spectators.

Lotteries

A common lottery format is that known as 6/49, as in the UK National Lottery. Here 49 rubber spheres, painted with different numbers, are whirled around a plastic tub, six are chosen at random. Gamblers pay £1 to select six numbers, and win a prize if their selection contains at least three of those winning numbers. But since only 50% of the takings go into the prize money, the mean return to Lottery players is far less than in casinos, or at the racetrack.

The main attraction is the prospect, however remote, of an enormous prize – one UK ticket has won over £22 million, and prizes in the USA have exceeded $300 million. Counting tells us that the probability of winning a share of the top prize from one ticket, by matching all the winning numbers, is about one in 14 million in the UK, less than one in 116 million in the Euromillions game, and around one in 176 million in the USA Mega Millions game.

To appreciate just how tiny these chances are, fix on the UK game. Figures show that the probability of death within one year for a random 40-year-old man is around one in a thousand. So the chance of his death within a day is about one in 365,000, within an hour it is about one in 9 million, so to get down to one in 14 million we are talking about the chance he dies within the next 35 minutes! For the Mega Millions game, under the same assumptions, the chance his ticket wins a Jackpot share is comparable to his chance of death within the next three minutes.

Despite the low return and forbidding odds, 'utility' gives a rational explanation for buying tickets. In exchange for £1, you will get back 50p on average anyway, and the other 50p buys you the right to dream about your future luxurious lifestyle, your philanthropy, and the possible envy of people like me who assured you it was a waste of money. These rights surely have some utility.

We shall assume that future lottery draws are independent of past results – an inanimate rubber sphere cannot 'remember' whether it is 'due' to be chosen. Short of cheating, there is no way of changing the chance of winning a prize. But you *can* influence the size of any prize, in those lotteries where a fixed proportion of the prize fund is shared out among the winners at each tier. There is an opportunity to exercise a little skill.

It arises because certain numbers, typically low (birth dates) and odd, are chosen more often than others, and because many lottery players spread their choices evenly over the ticket, perhaps in the mistaken belief that doing so means that they are selecting 'at random'. In consequence, combinations with several high numbers, or with numbers clustered together, or on the edge, are chosen less often. If you can identify the sort of choices other players are making, and *do something different*, your winning chance is not affected, but if you do win, you will win more than average.

Beware of trying to be too crafty, like selecting {1, 2, 3, 4, 5, 6}, or the winning numbers in the last draw, on the grounds that 'Nobody else will think of that'. They will. When the UK Lottery began, about 10,000 people were choosing the first six numbers. In September 2009, the winning numbers were identical in two successive draws in the Bulgarian Lottery: no-one chose them the first time, but 18 did so the second time.

Provided that other players continue to mark their tickets much as they have in the past, the following procedure, for UK-type 6/49 lotteries, will help you. Take an ordinary deck of 52 playing cards and discard three of them. Identify the remaining cards with the numbers 1 to 49, shuffle well, and select six cards. This is a way of choosing six numbers completely at random. Human beings cannot make such a selection unaided, they need this sort of auxiliary help.

And use these six numbers, *provided that*

(a) they total at least 177 (to give a bias to high numbers), and
(b) when marked on the ticket, they fall into two, three, four, or five clusters, and
(c) three, four, or five of them are on the outside border of the ticket, and
(d) they do not form any obvious pattern on the ticket.

If any of these conditions fail, return the six cards to the pack, shuffle it thoroughly, and repeat this sequence.

If you follow this recipe, you should still expect to lose money – the overall payback of only 50% of money received is hard to overcome. But you are less likely to have to share any Jackpot with the world and his wife.

TV games

Golden Balls was first aired in 2007. The last two players are faced with eleven spheres (the Golden Balls), some of which are worth money, the rest (the Killers) are worth zero. The players select five of these spheres to generate a potential prize fund; any Killer chosen reduces the value of any previously selected Ball to one tenth of its current value. Thus two Killers chosen after a £50,000 Ball make it worth just £500.

All the spheres are outwardly identical, so the players are choosing completely at random. There are 462 ways to choose five objects from eleven, so the chance of picking the five most valuable Balls is 1/462. In the first 288 shows, this occurred just once.

Take a Ball nominally worth £1,000: even ignoring the Killers, the chance of selecting it is only 5/11, so its real mean value is £455. Any Killers will reduce this sum even further – with three Killers, its mean value can be calculated as £255.

When the five Balls have been chosen, and the actual prize fund is known, each player makes a private decision, whether to *Split* the fund with the other player, or seek to *Steal* all of it. They reveal their choices simultaneously: if both Split, they share the fund, if just one of them Steals, that player gets the entire fund, if both Steal, neither get anything.

This scenario is well known in Game Theory, under the title 'The Prisoners' Dilemma'. Suppose your opponent chooses Split: then you are better off if you Steal. And if the opponent chooses Steal, you won't get anything anyway. So whatever choice the other makes, you can argue that choosing Steal will never lose. Frequently, both select Steal, and the only winner is the TV production company who pay out zero.

Versions of *Deal or No Deal* have been shown in over seventy countries. In the UK, there are 22 sealed boxes containing different amounts ranging from 1p to £250,000. The boxes are allocated randomly to 22 players, one of whom, Amy, will play that day. Her own box remains closed until the end. She first selects five other boxes, whose contents are revealed. A banker then offers a sum of money in exchange for the amount in her box. To accept this, she says 'Deal', ending the game, while the words 'No Deal' reject this offer. If the game continues, more boxes are opened, a new offer is made, and so on.

At the time of any offer, the exact amounts still in play are known, so their mean is easily calculated. In the early rounds, the banker's offer is normally far less than this amount, but Amy must have her utility function firmly in sight: if she strongly desires £5,000, and the offer is £5,400, she could rationally accept, even if the mean amount in the boxes left is over £20,000 – she might end up with 1p if she hangs on.

One time in 22, Amy will own the box with the top prize, but she will win that amount far less often. Utilities give a convincing explanation. At the final decision point, two boxes remain, one with £250,000 and the other with maybe £2. If the banker offers £80,000, even though this is well below the mean value of £125,001, only the bravest or richest Amy will reject it. A bird in the hand....

Provided the banker always offers less than the mean amount in the remaining boxes, the Law of Large Numbers ensures that, in the long run, contestants who Deal take home less than the amount in their box. So a real bank, that did pay out the offer but also received the amount in the box when the contestant Deals, would make a long-run profit.

The Colour of Money was billed as the most stressful show on TV, yet it survived only a few episodes in 2009. But it does give a splendid opportunity to illustrate uses of the Addition and Multiplication Laws in finding a probability.

The sums £1,000, £2,000,...,£20,000 were randomly allocated to twenty boxes of different colours, and the player, Paula, sought to reach some pre-assigned target, say £64,000. To do so, she could select up to ten boxes, one at a time. If she (unknowingly) chose the £14,000 box, the amounts £1,000, £2,000,... up to £14,000 would appear in that order at a stately pace: she could call Stop at any stage. If she made that call in time, she banked the amount last showing, but if she waited too long, she banked nothing. If, after ten boxes, she had not reached the target, she won zero. What tactics should she use?

Colour apart, all the boxes are identical, so Paula makes a completely random selection from those left in each round. In her last round, with eleven boxes left, her strategy will be obvious: for example, if she needs a further £9,000 to hit her target and exactly six boxes are worth £9,000 or more, she will hope to call Stop when £9,000 is shown, and her winning chance is 6/11. But what should she do in earlier rounds?

Perhaps the twelve boxes left with two rounds to go contain (in units of £1,000) the amounts 1, 4, 5, 6, 9, 10, 12, 13, 15, 17, 19, 20, and she requires another £15,000. It makes no sense to call Stop when she sees £7,000; if that figure ever appears, she knows that her box contains at least £9,000, so she could Stop at that sum, plainly better tactics. She can restrict her options to selecting from the twelve values in the boxes. The same argument also applies at the earlier rounds – her best call of Stop will always be at a value corresponding to one of the remaining boxes.

If Paula does intend to Stop at £9,000 here, she can argue: 'Eight of the twelve boxes have at least that amount, so my chance of success is 8/12. And if I do succeed, I'll need just £6,000 in the final round, and then eight of the last eleven boxes will work. The Multiplication Law tells me that chance of both of these is (8/12)*(8/11) = 64/132. Also four boxes have less than £9,000, so the chance I bank nothing first go is 4/12; I then need £15,000 from the last box, with chance 4/11. By the Multiplication Law again, the chance this path will work is (4/12)*(4/11) = 16/132. These ways of winning are disjoint, so the Addition Law gives the overall chance of success as 80/132.'

She can make a similar analysis for her other choices, such as going for £6,000, or £12,000. I invite you to do the sums – the Appendix describes her best choice.

In the planning stages of this show, the idea of using an expert mathematician to offer running advice was mooted. Paula could suggest she will try to call Stop at £8,000, the expert might say 'Not a bad choice. If you do that, you've got a 75% chance of winning the money. But if you plan to Stop at £11,000, your chance goes up to 80%.'

You can well imagine what could happen! Everything the expert said was correct, Paula changed her call – and failed to win the money, while her original instinct would have worked. Some tabloid newspaper would surely scream 'Maths Boffin Robs Army Hero's Widow of £64,000'.

All of us who investigate the maths of TV game shows are relieved that no such mathematical advice was ever given on this show!

Card games

The Law of Large Numbers means that you will receive your fair share of good or bad cards in the long run, so differences in skill levels will tell eventually. We look at three popular games.

In Blackjack, the house must follow fixed rules about when to draw cards, the player can do what he likes. Until Edward Thorp started winning significant sums, casinos believed that no system could beat their built-in advantage. Their logic had a fatal flaw: although they could expect to win 1–2% of stakes with a full stack of six or eight decks of cards, after a few deals the odds might shift in favour of the player. The casinos had omitted to use the *conditional probabilities* based on what cards remained unused, rather than rely on the probabilities calculated for a full stack.

Thorp developed a way of keeping track of which cards were left in the stack. When there is a high proportion of high value cards remaining, it becomes more likely that the rules compel the house to draw a card that leads to a losing total of above twenty-one: in the same circumstances, the player can opt not to draw. Thorp would bet the minimum amount so long as the stack had a low or average proportion of high value cards, then larger amounts should the stack composition shift in his favour. Simple, but effective.

When the stack composition does give an advantage to the player, how much should he bet? John Kelly had answered precisely that question a few years before Thorp's analysis: he should bet that proportion of his capital that is equal to the size of his advantage. This choice maximizes the rate at which his capital will grow.

For example, suppose he has £1,000 and the game is slightly in his favour – his winning chance is 51%, his losing chance is 49%. His advantage is 2%, so he bets 2% of his current capital, i.e. £20. Next time, he will have either £980 or £1,020, so if his advantage remains at 2%, his bet will be either £19.60 or £20.40, according to which outcome pertained. If he were too greedy – betting 10% of his capital when Kelly indicates just 2% – then he would eventually be ruined, despite his advantage. His capital is finite, and the stake would be too high to stand the inevitable losing streak.

Casinos take steps to identify and ban proficient card counters. No finer tribute to the power of understanding probability has ever been paid.

We noted that Bayes' Rule is the proper way to see how pieces of evidence should change our beliefs about Guilt or Innocence in a court case. In card games like Whist or Bridge, using this Rule can improve your chances of making the best decisions during play. For convenience, I retain the legal vocabulary, and use the word Guilty to mean that a particular opponent holds certain cards, say both King and Queen of Hearts, while Innocent means that she holds at most one of those cards.

By counting, we can find the proportion of all possible deals where she holds both cards, to give an initial assessment of the probability of 'Guilt'. It turns out best to convert this probability into its equivalent odds, in the standard manner. As this calculation is made at the outset, we say that we have found the *Prior odds* (of Guilt).

As cards are played, relevant *Evidence* emerges – perhaps she plays the King of Hearts on a trick. To see how such Evidence affects the odds of Guilt, a quantity termed the *Likelihood Ratio* is found: first, assess the probability of the Evidence assuming Guilt (she holds both King and Queen), then find its probability assuming Innocence (she has at most one of them). The Likelihood Ratio is just the ratio of the first to the second.

We can now deduce the *Posterior odds*, i.e. the odds of Guilt, taking account of this Evidence, using Bayes' Rule which says

Posterior odds = Prior odds × Likelihood Ratio.

Its format is plainly sensible: the bigger the Likelihood Ratio (i.e. the more likely is the Evidence when the opponent is Guilty), the more the odds of Guilt increase – but this Rule tells you *precisely* how much the Evidence affects the chances of Guilt.

To see it in operation, consider a realistic situation: our opponent either holds just the King and Queen (Guilt), or she has the King only (Innocence); the Prior odds are that those alternatives are just about equally likely. If she is Guilty you do best to play the Ace, if she is Innocent you should play some other card. Evidence now appears – she plays the King.

Without the Evidence, you must guess, and you will make the winning play half the time. With Innocence (she has King alone), the probability of the Evidence (she played the King) is 100%; but with Guilt (she had both King and Queen), she might equally well have played the Queen rather than the King that you saw, so the probability of the Evidence is only 50%. Their ratio is one half, so the Rule tells you that the Posterior odds are one half, i.e. she is twice as likely to be Innocent as Guilty – she is twice as likely to have the King alone. Not playing the Ace is the right decision two thirds of the time.

If, by this proper use of probability, you will make the winning play two thirds of the time, rather than just half the time, you should expect to do much better. You cannot guarantee to make the winning play, but you can improve your chances of doing so.

Bridge players refer to this scenario as the Principle of Restricted Choice – if the opponent had King alone she had to play it, with both King and Queen she had a choice. The fact that she did play the King shifts the odds towards her having to do so.

Today, the most popular form of *Poker* is Texas Hold'Em. Each player is dealt two cards and seeks to make the best poker hand possible from her own cards, and five communal cards that are dealt face up later. Which of the following hands is best, in the sense of being more likely to beat either of the other two when those communal cards are dealt?

Hand A: Two of Clubs, Two of Spades
Hand B: Ace of Spades, King of Diamonds
Hand C: Jack and Ten of Hearts.

Trick question, of course: after careful counting, it turns out that Hand A will beat B about 52% of the time, B beats C 59% of the time, while the chance C beats A is around 53%. So you would rather hold A than B, and rather hold B than C, but also you prefer C to A! Your chance of winning exceeds 50% if you let your opponent pick any of the three hands, provided you may then choose either of the others for yourself.

There is far more to poker than facility with probabilities. You must make judgements about what cards your opponents are likely to hold, and when you might bluff. But sometimes probability is very useful. Suppose the pot has 50 chips, and one communal card remains to be dealt. You can see that, if this final card is a Spade, you will make a Flush, which must win; if it is not a Spade, an opponent will win. Should you bet more chips to remain in the game?

Ignore how much you have already put in the pot. All that matters is the future. You can see six cards – two in your hand, four communal cards on the table. Of the 46 unknown cards, nine are Spades that give you victory, the rest lead to defeat. With 50 chips already in the pot, is it worth paying 10 more to see the final card dealt? 20 more?

By working out the mean profit (or loss) if you must pay x chips to see the final card, find the cut-off value of x that will, on average, give a profit. The Appendix gives the answer.

Chapter 7
Applications in science, medicine, and operations research

We may assess or interpret probabilities in different ways according to the context. But, as David Hand wrote in his *Statistics: A Very Short Introduction*, '... the calculus is the same', i.e. how probabilities are manipulated does not change.

Keep in mind the central ideas of the subject: the Addition and Multiplication Laws; independence; the Laws of Large Numbers linking frequencies to objective probabilities; Gaussian distributions when summing random quantities; other frequently arising distributions; means and variances as useful summaries. We may not expect our knowledge of the relevant probabilities to have the precision available for the examples in the previous chapter, but an approximate answer to the right question can be a reliable guide to good decisions. As statistician George Box said: 'All models are wrong, but some are useful.'

The next two chapters illustrate applications, loosely grouped under the chapter titles.

Brownian motion, and random walks

In 1827, the botanist Robert Brown observed that pollen particles suspended in liquid move around, apparently at random. Nearly eighty years later, Albert Einstein gave an explanation: the particles were constantly being buffeted by the molecules in the liquid. This movement is, of course, in three dimensions, but to build a satisfactory model, we first consider movement just along a straight line.

Suppose that each step is a jump of some fixed distance, sometimes left and sometimes right, independently each time. This notion is termed a *random walk*. The position after many jumps depends only on the difference between the numbers of jumps in each direction; the mean and variance of the distance from the start point are proportional to the number of jumps made.

Make a delicate computation: over a fixed time period, *increase* the frequency of the jumps, and *decrease* the distance jumped. With the correct balance between these two factors, the limit becomes continuous motion, the random distance moved having (by the Central Limit Theorem) a Gaussian distribution whose mean and variance are both proportional to the length of the time period. If movements left or right are equally likely, the mean will be zero.

Einstein's explanation for Brown's observations is that particles move in three dimensions, movement in each dimension following a Gaussian law for the reasons given. He made predictions about how atoms and molecules behave, provoking experiments that removed any lingering doubts about their existence.

The term 'Brownian motion' ought to be reserved for the actual movement of particles in a liquid, but it is also often used for this mathematical model of that movement.

Random numbers

The phrase 'random number' refers to one of two ideas. First, as in ideal games with dice or roulette wheels, one number from a finite list is chosen, all of them being equally likely. Second, as in the notion of snapping a stick at a random point, some point in a continuous interval is chosen, no part of that interval being favoured over another. The facility to choose long sequences of such random numbers, each value being independent of all the others, has many applications, as the next section will illustrate.

In 1955, a splendid book *One Million Random Digits* was published. It follows its title exactly: page after page of the digits zero to nine, grouped in blocks for ease of reading, but successive digits are entirely unpredictable – whatever the recent sequence of digits, you have one chance in ten of guessing the next one. Today, modern computers have built-in software to achieve the same ends. An initial value (the *seed*) is fed in, and a fixed mathematical formula produces the next value, which acts as a new seed, and so on. There is nothing random about this process at all, and if the same initial seed is used, the same sequence is generated. But, with a cunning choice of this mathematical formula, the sequence generated passes a battery of statistical tests and *looks*, to all intents and purposes, as though it were random. The term *pseudo-random sequence* is used.

No matter how much care is taken in this process, there will always be some lingering fear that hidden flaws in the method used will matter in the use to which the numbers are put. With that caveat, and relying on the experience of a large number of respected scientists, I am prepared to act as though my computer produces acceptable sequences of random numbers on demand. (The obvious danger of insider fraud means that these methods have no place in choosing numbers in Lotteries, or in UK Premium Bonds.)

Monte Carlo methods

How many different numbers will appear on 37 consecutive spins of a standard European roulette wheel? In theory, it could be anything between one and 37, but those extremes would occur very rarely; what is the most likely number of different numbers?

When this problem was first put to me, I did not immediately see an easy way to solve it. There are 37^{37} (a number with 59 decimal digits) possible outcomes of spinning the wheel 37 times, and when you try to write down all the ways in which, say, 28 different numbers could arise, you quickly lose enthusiasm. A more appealing approach was to perform a so-called *Monte Carlo simulation.*

Here, the computer's stream of random numbers was used to simulate the outcomes of 37 spins of a wheel, after which the computer counted how many different numbers had arisen. This process was repeated one million times, leading to 24 different numbers on 203,739 occasions, while 23 arose just 199,262 times. The nearest rivals, 22 or 25 numbers, each happened fewer than 160,000 times. The Law of Large Numbers says that the frequencies of the different outcomes will settle down to their respective probabilities, and these figures essentially settled the matter: the most likely result is that 24 different numbers will arise, and the chance of this is just over 20%.

Days later, I kicked myself for not spotting a standard way to solve the problem! I could calculate the exact probability of getting X different numbers in 37 spins, for any value of X, confirming the conclusion described above. But this does not invalidate the use of simulation to attack this sort of problem – quick and dirty answers can be useful. Indeed, the fact that the simulation gave answers consistent with the exact calculation boosted my general faith that the computer's random number generator was behaving as intended.

A more serious use of Monte Carlo methods occurs in polymer chemistry. A molecule consists of a large number of atoms, connected along a randomly twisting chain. Atoms can occur only at places on an evenly spaced lattice and, crucially, no two atoms can be in the same place. How far is it likely to be from one end of the molecule to the other?

We can think of the atoms as being at the places visited by some drunkard, staggering around at random on a three-dimensional lattice for a while, but somehow never visiting the same place twice. Without the requirement that no place be revisited, mathematical experts can make good progress, but that restriction seems to complicate the problem beyond theoretical attack. However, even a semi-competent computer programmer can write a sensible simulation of this complex, twisting, chain, and, by making one million, ten million, even a billion repetitions, obtain an answer as precise as is required. (Recall de Moivre's work, that precision increases only as the *square root* of the size of the simulation.)

Or suppose you want to estimate the area of an irregularly shaped leaf. Draw a rectangle around the leaf, and then simulate the positions of a large number of points scattered at random within that rectangle. Your estimate comes from multiplying the area of the whole rectangle by the proportion of points that fall within the leaf's boundaries.

As a final illustrative application, suppose Paul is hoping to set up a new petrol filling station. If he installs four pumps, the minimum viable number, there will be room for up to eight other cars to wait in a queue; each extra pump removes two waiting spaces, so if he installed the maximum of eight pumps, there would be nowhere to wait. To work out how many pumps will maximize his profits, he can carry out simulations of what would happen if he installed four, five, six, seven, or eight pumps.

As well as the installation costs, the running costs, and the profit margins, he would need to know the rate of arrival of potential customers and the distribution of the time it takes from a car pulling up at a pump until the pump becomes free. He should also take account of the chance that a potential customer would drive past if no pump were free, or the queue too long. All these figures are relatively easy to find or estimate, and it is *far cheaper* to make computer simulations than to experiment physically with various numbers of pumps over several months. Since he can use the same initial seed each time, he can run all his simulations under precisely the same conditions, thereby improving the comparison between the different estimates of profits.

Why 'Monte Carlo'? Despite the obvious connection between random numbers and casino games, the name was originally just a code word to refer to the use of these methods in military matters, including the early development of nuclear bombs.

Errors in codes

Morse code demonstrates how to transmit messages using only two symbols, 0 and 1 say. But some symbols may get corrupted, so that an initial 0 arrives as a 1, and vice versa. Even with a low error rate, the message received could mean something quite different from that sent. How best to deal with this?

Suppose each symbol sent has, independently, some small probability of being corrupted. We could repeat the symbols, but a moment's thought shows that sending 00 and 11 rather than 0 and 1 respectively is no use at all: if 01 or 10 arrives, it is a pure guess as to whether 00 or 11 had been sent. We'll guess correctly half the time, but doubling up on the symbols sent means that we should expect twice as many errors, so these factors largely cancel out. But consider transmitting 000 or 111 instead of 0 or 1.

Using 'majority vote' to decode messages, all of {000, 100, 010, 001} will be interpreted as 0, the other four possibilities will be interpreted as 1. If just 1% of symbols sent are corrupted, then when 000 is sent, using the binomial distribution shows that there is a 99.97% chance that one of those four sequences above arrives. That means that the error rate drops from 1% to 0.03%, a factor of over thirty. We would do even better if each digit were repeated five times, but at the expense of the message length. The best choice will depend on the size of the inherent error rate, and the speed of transmission.

Amniocentesis

Prospective parents (and statisticians) Juanjuan Fan and Richard Levine considered whether or not Fan should undergo an amniocentesis – a test to see whether the foetus she was carrying might have Down's Syndrome. Their experience can act as a template for others in a similar situation.

Knowledge of Fan's age, and a simple blood test, put the chance of Down's as 1 in 80; an ultrasound image was encouraging, leading to a use of Bayes' Rule that reduced the chance to 1 in 120. Amniocentesis is an invasive test – a hollow needle is inserted into the abdomen to extract a sample of amniotic fluid; if the extra copy on chromosome 21 that is characteristic of Down's is present, it will certainly be detected, but the test has a risk, estimated in this case at 1 in 200, of causing a miscarriage. Should parents, who would choose an abortion if Down's were present, take this test?

Fan and Levine reached their decision by the logical process of maximizing their expected utility. The worst possible outcome, miscarriage of a foetus without Down's, was assigned utility zero, the best outcome, birth without Down's, was given the utility of unity. To give birth to a Down's child, having opted out of amniocentesis, was allocated utility x, while to take the test and find Down's has a somewhat greater utility, y. (The possibility of a miscarriage is irrelevant in this last case, as the foetus would be aborted anyway.)

The expected utilities with and without the test are made. The test should be taken if the first exceeds the second which, in this case, reduces to requiring that y should exceed (119/200)+x; in round figures, y is bigger than 0.6+x.

If Fan and Levine had felt that the utility of discovering the presence of Down's, and thus having an abortion, was below 0.6, there would never be any point in taking the test. And the higher the utility they attached to having a child, albeit with Down's, the higher that threshold would become. If that utility were above 0.4, they should never take the test.

Choosing appropriate values for x and y requires some thought; and if the basic parameters – a 1 in 200 chance of a miscarriage through the test, a 1 in 120 chance of Down's without the test – were different, the final criterion would change. (See the Appendix.) Plainly, if the chance of Down's were less than the chance of a miscarriage, it would never be rational to take the test (yes?).

Fan and Levene discussed their dilemma, and their agreed choice of utilities led them to opt for the test. There was a happy ending: no extra chromosome, and no miscarriage.

Haemophilia

Haemophilia is the general name given to a group of disorders that prevent blood from clotting if a cut occurs. The clotting agent is found on the X chromosome, and the chance it is missing is under 1 in 5,000. As females have two X chromosomes, they would suffer from haemophilia only if it is absent from both copies, giving a chance of under 1 in 25 million, but males possess only one X (and one Y chromosome), so almost all instances of the disease are in males.

If males do have haemophilia, this becomes known well before they have a chance to father children, but a female may unknowingly have one normal X and one without the clotting agent. Such females are termed carriers, and the chance they pass on the affected gene to any child is 50%. A daughter who receives this affected gene becomes a carrier, a son will suffer from haemophilia. That Queen Victoria was a carrier is certain, as her son Leopold was a haemophiliac, and at least two of her five daughters were carriers. She had three other sons who did not suffer that disease.

Suppose Betty has a brother who suffers from this disease; Betty has several children, including Anne. What is the chance that Anne is a carrier?

To answer this question, it is enough to find the probability that Betty is a carrier; Anne's chance will always be half that figure. We know that Betty's mother is a carrier, as Betty's brother is a haemophiliac. From this information alone, the probability Betty is a carrier is 50%. And if *any* of Anne's brothers have the disease, Betty would be certain to be a carrier, so we look at the case when all Anne's brothers are healthy.

This situation is tailored for Bayes' Rule. Let Guilt mean that Betty is a carrier: since the initial probability of Guilt is 50%, the prior odds are unity. If Betty is Innocent (not a carrier), the probability of the Evidence (all brothers healthy) is plainly 100%. But if Betty is Guilty, each brother has, independently, a 50%

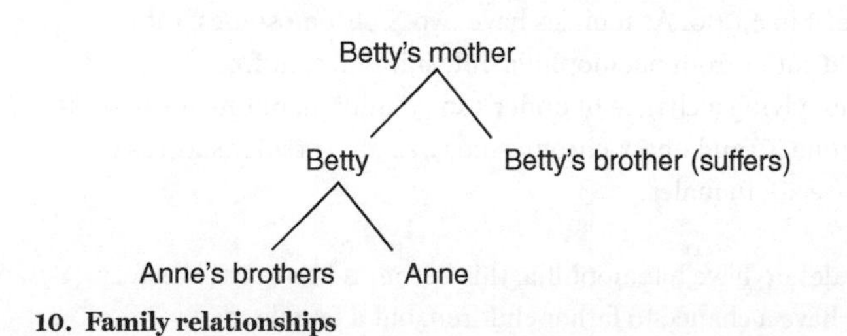

10. Family relationships

chance of escaping the faulty gene, so each healthy brother halves the Likelihood Ratio. Turning the posterior odds into probabilities, the chance Betty is a carrier is successively 1/3, 1/5, 1/9, 1/17, . . ., according as she has one, two, three, four, . . . brothers, all healthy.

For your own entertainment, suppose that Anne also has sisters, some of whom have sons, and none of these nephews of Anne are sufferers. How does that affect the chance Anne is a carrier? Check your answer against that given in the Appendix.

Epidemics

The phrase *herd immunity* expresses the fact that if sufficiently many people are vaccinated, then should a trace of infection enter the community, no epidemic will occur. Thus even those who were not vaccinated are very unlikely to catch the disease. Why should this be so, and how do we discover what 'sufficiently many' means?

Typically, those infected may transmit the disease to others, but those who recover have acquired immunity. So we label people as one of Susceptible, Infected, or Removed. The latter are those immune from infection because of vaccination, recovery, physical isolation, or death. It may seem callous to give these four outcomes the same word, but the brutal truth is that, so far as the spread of the epidemic is concerned, they really are equivalent!

To see how an epidemic might develop, let S be the number susceptible and I the number infected. Multiplying these two numbers together gives the total number of possible encounters between an Infected and a Susceptible. Crowded urban populations will be in close contact more often than scattered rural populations of the same size, and the probability of the Susceptible becoming Infected through an encounter will depend on the infectiousness of the disease. Overall, the probability that, during a tiny time period, some Susceptible individual becomes Infected will be of the form $\beta*S*I$, where β depends on both the infectiousness and how much people mix.

In this same tiny time period, any Infected person may move into the Removed category. Hence the probability that Infecteds reduce by one member will be proportional to the number infected, so takes the form $\gamma*I$, where γ depends on how fast those infected recover, are isolated, or die.

We have found the respective chances of an increase, or a decrease, in the number Infected. The balance between these two probabilities, $\beta*S*I$ and $\gamma*I$, determines whether an epidemic occurs. There is a good analogy with gambling. If the game is loaded against you, each bet is more likely to reduce your capital than increase it: your capital follows a random walk, with an inexorable drift towards zero. But if the game favours you, provided that bad luck does not bankrupt you early, the random walk ushers you far enough from zero to outrun any losing streak. To win a large sum, it is necessary, but not sufficient, for the game to be in your favour.

In the epidemic context, this means that the only time an epidemic (= large fortune) might occur is when any change in the number infected is more likely to be an increase than a decrease. In symbols, $\beta*S*I$ must exceed $\gamma*I$, which is the same as asking that S, the number of Susceptibles, exceeds the ratio γ/β, a quantity termed the *threshold* for this population. This is exactly what we were looking for: even if the disease enters into a population,

> an epidemic can only occur when the number susceptible exceeds this threshold.

William Kermack and Anderson McKendrick presented this result in 1927. Epidemics can be avoided by keeping the number susceptible below the threshold, which can be achieved in two distinct ways. First, vaccinate to reduce the number susceptible. Second, find means to increase the threshold. Being a ratio, a threshold increases when the numerator increases – we speed up recovery rates, or isolate infected people faster – or when the denominator decreases – we may be able to reduce the infectiousness, or we can ensure that people mix less, by closing schools temporarily, or postponing sports events where large crowds would gather. We can also assess the likely size of the benefits of these different responses, and so judge which are worth pursuing.

The same principles apply to controlling epidemics in animals. The first step taken to end a foot-and-mouth outbreak in cattle is usually to restrict cattle movements, which increases the threshold size by reducing the denominator. This is often accompanied by mass slaughter (not available with diseases in humans!), which not only reduces the size of the susceptible population, but also increases the numerator in the threshold.

This analysis also explains why we should expect epidemics of childhood diseases at fairly regular intervals. After an epidemic reduces the size of the susceptible population to below the threshold, new births, with insufficient vaccination, gradually take the size above the threshold, so creating the conditions for the next outbreak. The longer the interval between epidemics, the higher the vulnerable population, and the more severe will any epidemic be.

Knowing about probabilities may not cure diseases, but it can help mitigate their effects.

Batch testing

The army wishes to identify which of 1,000 potential recruits may be vulnerable to a certain disease, and thus unfit to serve. A blood test is available, but it costs £50 each time. Can the job be done for less than £50,000?

Provided that only a fairly small proportion will prove vulnerable, the answer is 'yes'. Choose some number K, and pool blood samples from K recruits; then test this pooled sample. If the result is negative, then *all* those who contributed blood are clear, and need no more testing; otherwise, with a positive result, at least one person in the group would test positive, so we use K more tests, one for each of them, to settle the matter. If we are lucky, one test will suffice, but we might have to make K+1 tests. We hope to make fewer tests than if each person is tested individually.

The best choice of how many samples to pool depends on the probability of a positive test. Suppose this is 1%. Then if we pool ten samples, there will be a positive sample among them about 10% of the time, while the chance all are negative is about 90%. So we will need only one test about 90% of the time, but eleven tests 10% of the time. That leads to about two tests on average. Pooling ten samples reduces the mean cost for those ten recruits from £500 to £100. If we split the 1,000 recruits into 100 groups of size ten, and pool their samples, we expect to save 80% of the initial estimate of £50,000!

More refined calculations show that, when the probability of a positive test is indeed 1%, we would do slightly better pooling groups of eleven, rather than ten, but the difference is very marginal. However, the best choice of the size of a pooling group *is* quite sensitive to the probability of a positive test. If we expect 2% of recruits to test positive, costs are minimized if we pool eight blood samples; with 5% the best choice is to pool five samples, while if 10% will test positive, pooling four samples turns out best. (Once again, use of the binomial distribution led to these answers.)

In World War II, this simple idea saved the USA about 80% of its initial expected costs.

Airline overbooking

Even though they must compensate passengers who are bumped off flights for which they have paid, airlines routinely sell more tickets than a plane's capacity. Simple economics is the reason: the cost to the airline of making the flight hardly changes with the number of passengers, but each empty seat is lost revenue. Expecting that not everyone who has booked a particular flight will show up, how does an airline work out the best amount of overbooking?

Suppose that the plane has one hundred seats, at a fare of £100 per seat, but a passenger is paid £200 if they have to be turned away because the flight is full. The airline needs a good estimate of the probability that a passenger who has booked will actually turn up. For charter flights to holiday destinations, this probability will be close to 100%, but it will be rather lower for passengers who have more flexibility in their travel plans. Frequency data will help airlines to estimate these chances.

Perhaps each passenger who books has, independently, an 80% chance of turning up for the flight. If 120 tickets are sold, total revenue is £12,000, and, although only 96 passengers will turn up on average, there is a chance, around 15%, that more than 100 show up, and at least one passenger must be left behind. (These numbers again come from using the binomial distribution.) The mean amount to be refunded because of overbooking in this case turns out to be £80. Selling five more tickets raises the revenue by £500, and the total mean refund increases by only £275, so this policy is expected to be more profitable. The most profitable policy, on average, comes from selling 128 tickets – compared with selling 125, the extra revenue of £300 just outweighs the mean extra refund cost of £295, while 129 tickets would be slightly worse than this.

It is more realistic to suppose that some passengers are more likely to turn up than others, and that groups of passengers who book together will either all turn up, or none of them will. But these details can be incorporated into the model, and airlines will continue to sell seats until the expected cost of compensation exceeds the extra revenue.

Queues

One of the best-developed applications of probability is the study of queues of various kinds. The initial impetus came from attempts to understand congestion in telephone lines – the work of the Danish telephone engineer Agner Erlang is honoured by the use of his name as the unit of the volume of telephone traffic. Queuing theory contributed to the success of the Berlin airlift of 1948/9, and the systematic study of queues flourished during the next twenty years.

David Kendall introduced notation, with the format A/B/n, now universally accepted as a shorthand way of describing queues where customers arrive singly. The first component, A, refers to the distribution of the time between customer arrivals, while B describes the distribution of the time it takes to serve a customer, and n is the number of servers.

For example, in the expression D/D/3, 'D' is short for deterministic, meaning that there is no randomness at all. Customers arrive on the dot at fixed time intervals, all service times also have exactly the same length, and there are three servers. This queue would be of little interest in the field of probability, as there is no variability. But suppose there is a huge number of potential customers, each of whom has a tiny chance of turning up in a given short time interval, so that customers arrive at some overall average rate, but completely at random. Here the symbol M is used, to honour Andrey Markov, so M/D/2 means that customers arrive at random, and select either of two servers, who each take some fixed time to do their job.

We want to know how queues will behave. The main questions are how long do customers have to wait, how frequently are servers just sitting on their thumbs, and what can we do to improve matters? The 'servers' may be the intensive care beds, while 'customers' are patients needing that care.

If customers arrive every five minutes on average, and there are three servers, then unless the mean service time is less than fifteen minutes, a queue of indefinite length will build up, and the whole operation is unsustainable. So we must assume that the mean service time, taking account of the number of servers, is less than the mean time between arrivals. The ratio of these two mean times is termed the *traffic intensity*, some figure between zero and unity.

In an ideal world, customers would face only short queues and the server would be busy nearly all the time. But these two requirements are diametrically opposed. Take a simple case with one server and customers arriving at random. If the traffic intensity were 0.9, calculations show that we might expect about five waiting customers on average, and an empty queue about 10% of the time. If the intensity rose to 0.98, the server would be unoccupied just 2% of the time, but the mean queue length would shoot up to 25. Most customers would consider this a worse arrangement. Unless servers have enough 'idle time', customers will get angry, leave the system, or do both.

The queue behaviour depends on much more than just the traffic intensity. Other things being equal, the more variable the service time, the longer you can expect the queue to be. With several servers, it matters whether they operate as in my railway station, where one central queue feeds up to six servers, or as in my supermarket, where I choose which aisle to join, and stay there. In some situations, e.g. ambulance calls, some customers may have higher priority. Many queues use the 'First come, first served' rule, but when non-perishable goods are stored on a shelf awaiting use,

‘Last come, first served’ may apply. Some queues feed other queues, the servers may work at different speeds, bunches of customers could arrive together. Eagle-eyed queuing theorists have found answers to the central questions under almost any realistic model you can think of.

Chapter 8
Other applications

The applications of probability away from dice, casino gambling, and various aspects of natural science can be overlooked. In this chapter, I have picked out some of its appearances in law, social science, sport, and economics to emphasize its ubiquity.

The common theme is that decisions we make will depend on the probabilities of various outcomes, so we need methods that lead to reasonably reliable estimates of those different probabilities.

Legal matters

Although Lord Denning, one of the best-known UK judges in the 20th century, had a mathematics degree, few lawyers feel comfortable with probability. This ought to be astonishing, as phrases relating to the subject are used freely in courts. In civil cases, such as libel, to say 'on the balance of probabilities' clearly puts the dividing line at 50%. But in criminal cases, where a jury is asked to convict only if they are 'sure' of Guilt, there is no consensus on a figure. Some people would wish to convict if they were 80% certain of Guilt, others would use 95% or even higher. These are plainly subjective probabilities. And although the same phrase is used whatever the offence, some would apply a lower threshold of proof for a relatively minor offence. This could make it harder to convict mass murderers than fare dodgers.

Suppose an expert witness testifies that the DNA of the accused matches DNA found at a crime scene, and that the chance of a match between the latter and an innocent person chosen at random is one in several million. Jurors may have two distinct problems with this statement. The first is that they may think that it is equivalent to saying that the chance the crime scene DNA is NOT that of the defendant is one in several million. The second is that they may treat all such tiny figures as equivalent, even though one in ten million differs from one in a billion by a factor of a hundred.

The first error has been termed 'The Prosecutor's Fallacy'. Starkly, it is equating the chance of Innocence, given a DNA match, to the chance of a DNA match, given Innocence. This is logical nonsense: the chance of zero arising, given a fair roulette wheel, is not the same as the chance that the wheel is fair, given that zero occurred. This trap can be avoided by giving the jury an estimate of how many citizens might match the crime scene DNA. With a population of around 60 million, if the match chance is one in 2 million, there might be thirty or so; if it is 1 in 20 million, there might be about three; it is unlikely there are more than half a dozen. But do not overlook the phrase 'selected at random': the more close relatives the criminal has, the more matches we would expect, the less strong this evidence against the accused.

The second error can best be avoided by remembering how Bayes' Rule measures the usefulness of any piece of evidence. Before this evidence is presented, you have some idea of the odds that the accused is Guilty. If the evidence is ten times more likely under Guilt than Innocence, then the odds of Guilt get multiplied by ten: while evidence that is three times more likely under Innocence than Guilt *reduces* the odds by a factor of three, and so on. With DNA evidence, it often happens that the chance of the evidence, assuming Guilt, is 100%, which makes the impact of the evidence clear: the odds of Guilt should be multiplied by whatever the 'several million' figure actually is.

Randomized response

A head teacher wishes to ascertain what proportion of his senior students smoke cannabis. Direct questions are unlikely to produce truthful answers, but a technique known as *randomized response* is available. The main idea is that the teacher recording the answers does not know what question is actually being asked, so that cannabis users are able to answer honestly without fear of being identified.

The words 'I smoke cannabis' are written on each of 80 cards, and 'I do not smoke cannabis' on another 20. Each card is placed in an identical envelope, these 100 envelopes are mixed thoroughly in a large bag. The students should see this operation being done, so that they know that the bag contains both versions of the question, and in those proportions.

Angela selects one envelope at random, opens it, reads the question to herself, and simply says either 'Agree' or 'Disagree'. She then puts the postcard back in the envelope, returns the envelope to the bag, and shakes the bag up ready for the next student.

Suppose that one-third of the responses are 'Agree'. Because the students are picking the envelopes at random, 'Agree' is the honest response from 80% of users, and 20% of non-users. A few lines of algebra show that this is consistent with 2/9 of the students being users. The head teacher has his answer, no individual student has been identified.

Alternatively, replace 'I do not smoke cannabis' with a question on an unrelated subject, for which the proportion of 'Agree' answers is known. If a previous survey has established that half the students own a pet, and there is no reason to link pet-owning with cannabis smoking, the statement on 20 cards could be 'I own a pet'. Then, if one-third of the responses are 'Agree', we estimate that 7/24 of the students are users.

The calculations giving these estimates are in the Appendix.

The uncertainty as to which question is being asked each time leads to some imprecision in the final estimate. The proportion of envelopes that contain the sensitive question should be as high as possible, but low enough for genuine cannabis users to believe that giving an honest answer will not have repercussions. Putting the sensitive question on as many as 95% of the postcards would not work.

WADA

The World Anti-Doping Agency seeks to promote sport as a healthy activity, by identifying athletes who take performance-enhancing drugs, and excluding them from competitions. But whatever methods are used, any testing programme is liable to two opposing types of error: claiming that an athlete is using drugs when they are innocent, and passing an athlete as clean when they are a drug user.

Unfortunately, methods of reducing the chance of making either type of error often tend to increase the chance of the other. For example, one test measures the ratio of testosterone to epitestosterone. The body normally produces these substances in fairly equal amounts, but those athletes who seek to cheat by injecting testosterone will have a high T/E ratio. Athletes whose ratio is above some specified amount, say six to one, will be banned. However, the T/E ratio varies naturally: it changes over a menstrual cycle, it will increase if you catch flu. Set the critical T/E ratio too high and no drug cheat will fail it; set it too low, and many innocent athletes will be wrongly accused.

Suppose the chance that a particular test makes a mistake is 1%. That means that if the athlete is innocent, the chance they fail is 1%, if they are users the chance they pass is also 1%. Sam fails the test: what is the chance she is innocent?

Put like that, the temptation to say '1%' looks overwhelming – this test gets things wrong one time in a hundred, so if it says she has failed, that will be wrong one time in a hundred. Resist this temptation. The only valid answer is 'We do not know. It could be any figure. We need to know the proportion of drug cheats in the population.'

For, suppose that proportion is 1% or so. Then among 10,000 athletes we expect 100 drug cheats, and 9,900 innocents. In testing, we expect just one drug cheat to pass, leaving 99 who fail. But 1% of the 9,900 innocents, i.e. another 99 athletes, will also fail the test. Among those who fail, *half* are innocent: the chance Sam is innocent would be 50%.

If the proportion of cheats differs from 1%, this conclusion changes. If it is higher, the chance that Sam is innocent will be less, but if it is lower, her chance of innocence will be even higher. The lower the proportion of drug cheats, the less satisfactory is this test, despite its apparently impressive performance.

This same logic applies when we consider how to detect potential terrorists at airports. Whatever screening devices are used, they cannot be perfect, but suppose that the probability a real terrorist evades these checks is tiny, 1/10,000, while the chance that an innocent person is led away for intensive interrogation is a minuscule 1/100,000. How likely is it that someone picked out is guilty?

We cannot answer the question without having some idea of the proportion of would-be passengers who are terrorists. Try one in a million – frighteningly high, given that Heathrow handles over fifty million passengers a year. But the figures assure us that, even with fifty potential terrorists, it is overwhelmingly likely that all will be detected.

Unfortunately, five hundred innocent passengers will also be detained! Among those stopped by this system, fewer than 10% are terrorists. And if there are fewer than fifty terrorists, the chance that someone who is stopped is indeed guilty is even lower. Detection methods must have much better performance figures if they are to be useful.

Football results (1)

Betting on the results of soccer matches generates a substantial interest in the UK. All sorts of exotic bets can be made – on the time of the first throw-in, the sum of the shirt numbers worn by all the goal-scorers, how many red and yellow cards will be flourished during the game – but most interest is on which of the three results, Home win, Draw, or Away win, will occur. A rational punter will assess the respective probabilities of these results, and his decision on whether to bet, and how much, will rest on these assessments as well as the payout prices offered by the bookies.

But how might the punter deduce his degrees of belief in the different outcomes? In May 2009, statistician David Spiegelhalter took up the challenge on BBC Radio's *More or Less* by analysing the ten games to be played in the Premier League two days later. For each game, he estimated the number of goals each team might score, on average, taking account its own strength in attack, and their opponents' defensive capabilities. For example, a strong Home team (Arsenal) were estimated to score 2.1 goals, on average, against Stoke City.

No team can score 2.1 goals, but that figure is just the average over a hypothetical number of matches. The crucial step is to assess the probabilities of 0, 1, 2, 3,… goals in a single game, and Spiegelhalter used the Poisson distribution. Data over many years show that this is pretty good at describing how the actual number of goals tends to vary around its average. With Arsenal's figure of 2.1, the chance of no goals came out as 12%, one goal as 26%, two goals as 27%, three as 19%, and so on.

Data for Stoke put their mean score as 0.67 goals. This translates into a 51% chance of no goals, a 34% chance of just one goal, 11% for two goals, and so on. With a leap of faith, take the numbers of goals scored by each team as independent. So the probability of a 2-1 score comes from multiplying the chance the Home team scores twice by the chance the Away team scores once – in this case, 27% * 34%, around 9%.

In this way, the probability of any possible score is estimated. Then the probabilities for each of Home win, Draw, and Away win are found from the Addition Law, by adding up the separate probabilities of all the scores that lead to those three respective results. This gave Arsenal a 72% chance of victory, Stoke had a 10% chance, leaving a chance of 18% for the Draw. The score given the highest probability, at 14%, was 2-0.

Do not scoff! In the ten games, the exact score given the highest probability happened twice, and eight of the ten match results were those that were identified as being the most likely. A betting man, who had placed money on each 'most likely result' and on each 'predicted' exact score, would have smiled happily as the match scores unfolded.

How can we reconcile a 72% degree of belief that Arsenal would win that match with ideas of frequency, as there is no question of playing this game hundreds of times and counting how often Arsenal won? Recall how we judged the reliability of a weather forecaster when she says that the chance of rain tomorrow is 30%: there is only one tomorrow, it will either rain or it will not. However, we can look at all the occasions when she gives rain a 30% chance, and check its actual frequency. We shall believe her claim about tomorrow, or not, on the basis of her overall record. With soccer matches, we can make similar calculations for all games played over the season. Among these, there might be forty or so where some result was given a probability close to 72% – we can check whether the 'predicted' result did occur with a frequency around 72%, as a way of validating our methods.

Can a gambler expect to make money by using these ideas? The payout prices depend heavily on how much is staked on each outcome, and the largest sums are usually staked on one team or the other to win. Bets on a Draw tend not to attract committed fans. If the chance of a Draw is assessed as 25%, and the payout price is better than three to one, the opportunity to profit is there.

Do not assume that the best bet is on the outcome with the highest predicted probability!

Football results (2)

Before the 2010 soccer World Cup Finals began, statistician Ian McHale published the results of his calculations, which allocated to each of the 32 teams some non-zero probability of winning the trophy. He made Spain the favourites, albeit with a winning chance of only 11.6%, followed by Brazil, whose chance was put at 10.3%.

To obtain these figures, McHale used an approach similar to that described above for each match. However, he did not make a direct calculation of the probabilities of the distinct match outcomes, he relied on a Monte Carlo simulation.

Thus, for a match in which England's mean score was put at 1.5 goals, the Poisson model gives a 22% chance of no goals, a 33% chance of one goal, and so on. The computer's random number generator selected one of the values 0, 1, 2, 3,... with the appropriate probabilities, and did the same thing for England's opponents, leading to some simulated score such as a 2-2 draw. Similar simulations were made for every scheduled match, leading to simulated group tables, and then to matches in the knockout stages all the way to the final. This process was repeated 100,000 times, and the number of simulations in which each team emerged as champions was recorded. Spain 'won' 11,633 times, hence the 11.6% figure noted earlier. The Law of Large Numbers, as usual, is the justification.

And Spain did win! Were McHale's probabilities 'correct'? We cannot know. Perhaps Spain would have won 65% of the time, had it been possible to make indefinite repetitions of the tournament. But the best evidence that his methods make good sense is that bookmakers follow a similar path to set their initial payout prices to attract punters.

Black-Scholes

Share prices on stock markets fluctuate, sometimes for no apparent reason. If the price is £5 today, you do not know what the price will be next month. However, you can buy an *option* – the right to buy (or sell) that share at the *strike price* of £5.20 at a given future time. If, at that time, the market price is less than £5.20, you will not exercise your option to buy, but if it is above that price, you can make an instant profit by taking up the option, and immediately selling. Corresponding remarks apply to an option to sell. What are fair prices for these options?

Fischer Black and Myron Scholes addressed this question in 1973. At the heart of their work was the assumption that the changes in share prices varied randomly, but in a particular way related to the Gaussian distribution. The fair prices for both buy and sell options were found to depend on the current price, the price at which the option would be exercised, the intervening time period, prevailing interest rates, and the volatility of the underlying share price (as measured by the standard deviation over a period): but *not* on the mean amount by which the share price was expected to change!

This last point may be surprising, but that is how things work out. It is also quite useful, as it means that we have no need to add to any uncertainty by estimating the trend in prices. If you want to discover the fair price for some particular option, free software is widely available – just type 'Black-Scholes' into your favourite search engine. Given the current price and the strike price, the fair cost of a buy option would increase if the time period were longer, or if interest rates were higher, or if the volatility of the share price were higher.

How do the claims in this last sentence accord with your intuition? The first does seem reasonable, as the longer you are prepared to wait, the higher the chance of an increase in the underlying share price, but the other two claims are more subtle. An increase in volatility also increases the chance of a jump in the share price, but it also happens to *decrease* the mean change in price – and the former effect turns out to be bigger.

The volatility is measured by looking at the changes in the share price over the 250 or so trading days in one year. This should give enough data for an estimate to be reliable, but not stretch so far into the past as to be irrelevant for current conditions. A poor estimate of the volatility will lead to an unreasonable price for an option.

A model is only useful when its key assumptions are not violated. And, as Figure 7 shows, taking the Gaussian distribution as a model for price fluctuations implies a really tiny probability for catastrophic events, such as the price dropping by more than three or four standard deviations. When the actual probability of such an event is significantly underestimated, the model is undermined, and the conclusions it indicates may have no sound basis at all. The extreme-value distributions, mentioned in Chapter 4, have been used to address this problem.

Share portfolios

Companies A and B are both expected to make profits. With low interest rates, A is expected to return 20%, while B should return 40%; with high interest rates, the positions are reversed – A should gain 40%, B 20%. Suppose Nick is a risk-averse investor, while Mary is risk-attracted.

If low or high interest rates are seen as equally likely, both companies may look equally attractive, with a mean return of 30%. In accordance with their respective attitudes to risk, Nick could divide his funds equally between the two companies, and guarantee to get 30%, whether rates are high or low: Mary could plump for one company or the other, hoping to get 40% but accepting she might get only 20%.

Suppose B is replaced by company C, which will return 10% with low interest rates, or 50% if rates are high – again an average of 30%, like company B. But now mixing A and C makes no sense to either investor: Nick prefers A alone, Mary puts everything into C.

The essential difference is that the returns from A and B are *negatively correlated* – in conditions when one is high, the other tends to be low; but returns for A and C are *positively correlated* – they do better or worse together. 'Correlation' is measured on a scale from −1 (total negative correlation) to +1 (total positive correlation). If two assets fluctuate in value independently of each other, their correlation will be zero.

Risk-averse investors are encouraged to diversify their holdings, so that any losses might be balanced by gains elsewhere. They wish to hold negatively correlated assets. But there is an inescapable piece of logic: if X is negatively correlated with Y, and Y is negatively correlated with Z, then X and Z will tend to be positively correlated!

However, all is not lost. A mathematical result, due to Salomon Bochner, proves that it is indeed *possible* for each pair of assets in a large portfolio to be negatively correlated; but the greater the number of assets, the harder it is to achieve mutual negative correlation.

Chapter 9
Curiosities and dilemmas

At the beginning of this book, I noted that some aspects of probability appear, at first sight, to defy common sense. Examples have arisen as the story has unfolded. Here are some other circumstances where intuition can be misleading, but, with sufficient care, these apparent contradictions can be explained. The subject of probability is wholly free from real paradoxes.

But although ideas of probability can help us make sensible decisions, we may also find that even thinking about the probabilities of certain events might lead to uncomfortable dilemmas.

Parrondo's paradox

Graham Greene's novel *Loser Takes All* is a fine read, but based on a false premise: that there is some clever mathematical way of combining bets on a roulette wheel to give the player an advantage, rather than the house. On the contrary: mathematics has proved that, when all individual bets favour the house, no combination whatsoever can turn matters round and favour the player. Sorry, folks.

Juan Parrondo has shown that you have to be very precise in how you formulate a general claim that, whenever all bets favour one side, it is impossible to combine bets so that the other side has an advantage. I describe here a modification of his idea, due to Dean Astumian, who described a simple game played on the board with five slots, shown in Figure 11. (This is not a *serious* game. It was constructed merely to make this point.)

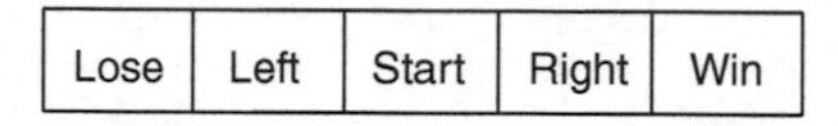

11. The board for Astumian's game

You need some way of generating a random event that will occur 1% of the time: perhaps a bag with 99 White balls and one Black ball, or a spinner that is equally likely to come to rest on any one of its one hundred sides. To begin the game, place a token on the slot marked 'Start'. Every move will take the token one step left, or one step right, and you win if the token reaches Win before it hits Lose.

There are two basic sets of rules, call them Andy and Bert. With Andy, from Start you always move to Left, and from Right, you always move to Win. From Left, you use the spinner, to give a 1% chance of moving to Lose, and a 99% chance of moving back to Start. With Bert, the spinner is used for Start to give a 99% chance of moving to Right, and a 1% chance of going to Left. From Right, you always return to Start, while from Left, it is the same as in Andy – the spinner gives a 1% chance of moving to Lose, a 99% chance of returning to Start.

Analysis of these games is simple. In Andy, there is no provision ever to reach Right; you shuffle around between Start and Left, until random chance takes you from Left to Lose. In Bert, you usually shuffle between Start and Right, with occasional visits to Left. Eventually, on one of these sojourns to Left, random chance takes you to Lose. In either game, the chance of reaching Win is zero.

For the new game, Chris, you also need a fair coin. At each turn, toss this coin: if it shows Heads, use the rules of Andy, if it shows Tails, use the rules of Bert.

It turns out that your winning chance in Chris exceeds 98%! It is easy to see why you are strong favourite: if ever you get to Left, you are overwhelmingly likely to return to the safety of Start. From Start, you play Bert half the time, with its 99% chance of getting to Right; and in Right, you play Andy half the time, inevitably winning.

Following either Andy or Bert, you *must* lose: flip between these games at random, and you win nearly every time! Framing a mathematical theorem that excludes examples like this, but confirms that Greene's plot rests on shaky ground, requires very precise language!

2+2=4, or 2+2=6?

Suppose we carry out Bernoulli trials with a fair coin, i.e. each toss, independently, is equally likely to be Heads or Tails. A typical outcome will be HHTHTTTHT.... The mean number of tosses until Heads appears is two; but what is the mean number of tosses until we see HT, or HH?

The intuitive answer is four, as we expect to wait two throws for the first symbol, then another two throws for the second. And the mean number of throws until we see HT is indeed four, but this is *not* the case for HH. To see that pattern, the mean number of throws is six!

The reason for the difference is that, to get HT, it is correct to argue that we expect to take two throws to get the H, then another two to get the T that completes the pattern. And Two plus Two equals Four. But for HH, after we have the first H, the next throw will be T half the time, and we must begin again – all throws up to that point will have been wasted. The algebra leading to the correct answer is in the Appendix.

Between H and T, each is equally likely to appear first; what about between HH and HT? Again, each is equally likely to arise before the other, since we must wait for the first Head, and then the next throw determines the answer. However, between HH and TH, the latter is three times as likely to appear first! The reason is simple: the sequence will begin with HH one-quarter of the time, but unless this happens, it is inevitable that TH appears first. (Think about it.)

The game *Penney-ante* is based on the above ideas. You invite your opponent to select any of the eight possible triples like HHT, or THT, etc., that might occur in three consecutive throws of a fair coin. You select a different one, a neutral person tosses the coin repeatedly, and the winner is the person whose triple is seen first.

Despite the apparent generosity of allowing your opponent to have first pick, this game favours you – if you know what you are doing. Whatever she chooses, you can select a triple that will appear before hers at least 2/3 of the time! The winning recipe is in the Appendix.

Give me a clue...

(1) Three double-sided cards of identical shape and size are placed in a bag. One card is Blue on both sides, one is Pink on both sides, the last is Pink on one side, Blue on the other. One card is selected at random, and one side of it is exposed, and seen to be Pink. Is the other side more likely to be Pink or Blue? Or are the chances equal? Over to you – answer below.

(2) Careful counting shows that a bridge hand of thirteen cards, dealt from a well-shuffled deck, will contain two or more Aces about 26% of the time. You deal out a hand to Lucy. To the question 'Does your hand contain at least one Ace?', she answers 'Yes'. On a

separate occasion, you deal a hand to Tina, and ask 'Does your hand contain the Ace of Spades?' She also responds 'Yes'. Which of the two hands is the more likely to contain two or more Aces? Or are the chances equal? See below.

(3) Suppose that, among 1,000 males and 1,000 females, all with satisfactory qualifications, 480 males but only 240 females gain admission to a university. A clear case of sex discrimination – men are twice as likely as women to be admitted?

The answers? With the Pink/Blue cards, seeing a Pink side plainly eliminates the double Blue card. All three cards were equally likely, just two are left, Pink/Pink or Pink/Blue. With one of these cards, the reverse side is Pink, but with the other card, the reverse side is Blue. It looks as though Pink and Blue are equally likely.

That reasoning is sloppy: Pink is twice as likely as Blue, and you can check this by doing this experiment a dozen or so times. Better, note that the cards have three Pink sides between them, and all of those are equally likely to be the side seen. But only one Pink side has Blue on the reverse – twice, a Pink side also has Pink on the reverse. (You could use Bayes' Rule, but that would be sledgehammer and nut territory.)

As an impoverished graduate student, Warren Weaver, one of the founders of *Information Theory*, taught other students the usefulness of understanding probability by consistently winning money from them when playing this game.

With the deck of cards, we know both times that the hand has at least one Ace, and many people would suggest that Tina and Lucy are equally likely to hold two or more Aces – all Aces are equally likely, so why should Tina confessing to the Spade Ace in particular make any difference? Reject those thoughts, and do the counting properly.

For Lucy, among the hands with at least one Ace, we find the proportion that have two or more – it is about 37%. For Tina, along with the Ace of Spades, she has twelve more cards, chosen at random from the remaining fifty-one. About 56% of the time, these include another Ace: Tina is far more likely than Lucy to have two or more Aces.

Your suspicious mind rightly tells you that the answer to the third question is 'No'. For suppose that, in the English department, 20% of 950 women who applied, and 10% of the 50 men, were admitted; in Business Studies, all 50 women gained admission, but only half the 950 men. Do the sums: 240 women and 480 men were successful, but, *in each department*, the success rate for women was twice that for men. Any discrimination was against men, not women!

Indeed, in real life, from thousands of applications to Berkeley's Graduate School, 44% of males but only 35% of females were admitted. However, when the applications were broken down into the separate departments, there was hardly any difference between the admission rates of men and women. But admission rates did vary between departments, and female applications were largest to those departments that admitted a smaller proportion from both sexes.

This counter-intuitive result is an example of *Simpson's Paradox*. It shows the perils of working with proportions, rather than absolute numbers, and it turns up all over the place.

It is far more than a mere curiosity. You have no justifiable claim to be numerate unless you know what it is about.

Do you want to know?

I have argued that probability is the key to making decisions under uncertainty, and I will not retreat from that. But the ability to know probabilities more precisely, and in new circumstances, throws up some uncomfortable dilemmas.

It is now possible for individuals to have their own entire genetic code sequenced, but Nobel Laureate James Watson and Harvard psychologist Steven Pinker have both opted not to know which version of a gene known as APOE they carry. Having one copy of the epsilon4 version of this gene quadruples the chance of developing Alzheimer's disease, while having two copies is associated with a twenty-fold increase in the chance. (Paradoxically, having this epsilon4 version is also associated with benefits during one's youth.) Another Nobel Laureate, Craig Ventner, knows that he does have one copy of epsilon4. One research laboratory has the policy of *never* disclosing to volunteers their APOE status, on the grounds that, with current knowledge, there is no treatment available to mitigate bad effects.

But some commercial companies may be *very* interested in your APOE status, indeed in your whole genome. If your genetic composition suggests a high probability of an early death, they may be willing to give greatly enhanced annuity rates – but may also demand much higher medical premiums. Companies who have full genetic information on an individual might 'offer' a bespoke service, tailored exactly to the life prospects of the client.

John and Tom are both 65 years old, and will each use £15,000 to buy an annuity; normal life expectancy is, say, 15 years, but John's genes suggest 10 years longer, Tom's 10 years less. Ignoring genetics, Company A offers both the same sum, £1,000 per year. But company B uses the genetic information, and offers Tom £3,000 per year, but John only £600 per year.

Recall the aphorism that, in the long run, averages rule. Both men would accept their higher offer, so company A should expect to pay out £25,000 to men like John, a loss of £10,000 each time, while company B expects to pay £15,000 to Tom and his ilk, and so break even. Company A will collapse, B will survive.

If the only viable insurance companies are those like B, we can expect many miserable people, who either cannot get medical/travel insurance at all, or who discover that their retirement plans are thrown askew because they are forecast to outlive their savings.

Barristers are advised to ask, in cross-examination, only those questions to which they already know the answers. Before you ask for your genome to be sequenced, assure yourself that you are fully prepared for what you might learn. Think of all the stages in life: a printout of your child's genome at birth may have devastating news; when contemplating marriage, should you and your betrothed take steps to learn the likelihood of any children being severely afflicted? Should your employer have the right to deny you promotion because you have an increased risk of some illness? Should candidates for high public office, say president or prime minister, have to disclose their genome, so that voters are more aware of any genetic propensity to become unstable?

A randomly selected UK female has a 12% chance of contracting breast cancer. But if she has inherited certain mutations of genes known as BRCA1 or BRCA2, the chance increases to 60%. Should a mother of three daughters, whose sister has this mutation, be tested herself? And if she is tested and gets bad news, at what ages (if ever) should her daughters be told that they each have a 50% chance of having inherited this mutation?

Whatever you feel in such uncomfortable circumstances, recall that it is 'probability', and not 'certainty'. If the probability of having the mutation is 10% for Emma, and 60% for Fiona, it may well turn out that Emma develops breast cancer, while Fiona does not. If they know their chances of having this mutation, they have to deal with this knowledge in their own way. To repeat the central dogma of decision theory: the rational action is the one that maximizes the mean utility of the consequences. You can never be sure you have taken the action that would have worked out best, but you have made optimal use of the information you have. You cannot ask for more.

Appendix

Answers to questions posed

Chapter 2: With fair dice having eight or ten sides, the chance of an even number is 1/2, and the chance of a multiple of six is 1/8 or 1/10 respectively. The respective chances of a multiple of three are 1/4 (two in eight), which gives independence, and 3/10, which does not.

Chapter 6: In the Colour of Money puzzle, you should aim to bank £12,000 at your first choice. If it works, you hope to bank £3,000 next time; if it fails, you must try for £15,000 next time. Your overall winning chance is (6/12)*(10/11)+(6/12)*(4/11)= 84/132=7/11, which is higher than any alternative tactic.

In 46 similar poker games, the nine times a Spade arises, you make a net profit of 50 chips; the other 37 times, you lose the extra stake. If you can stay in for 12 chips or fewer, you'll be ahead on average, but if you have to pay 13 chips or more, you'll expect to lose more than you gain.

Chapter 7: The chance of a miscarriage is 1 in m, the chance of a Down's baby is 1 in n, both m and n large and $m>n$. Take the test if $y>x+n/m$.

Suppose Anne's sister Celia has n sons, all healthy. If Betty is not a carrier, all Celia's sons are sure to be healthy; if Betty is a carrier, the chance Celia will not be a carrier (and hence all her sons are healthy) is 1/2, and the chance she is a carrier, but nevertheless all her sons are healthy, is $(1/2)^{n+1}$. So the posterior odds of Betty being a carrier, after the information about all Anne's healthy nephews, are the prior odds (i.e. those computed using the information about Anne's healthy brothers), multiplied by all these factors of the form $(1/2+(1/2)^{n+1})$ from her sisters. Convert this figure into the probability that Betty is a carrier, and halve it to find the chance Anne is a carrier.

Chapter 8: In the randomized response examples, suppose the proportion of smokers is x, and all answers are truthful. Then, in the first case, the overall proportion of 'Agree' answers is $0.8x+0.2(1-x)$; equate this to 1/3, solve to find $x=2/9$. In the second case, the proportion of 'Yes' answers is $0.8x+0.2/2$; setting this equal to 1/3 gives $x=7/24$.

Chapter 9: Let x be the mean number of throws to get HH. To get the first H takes two throws, on average. After this, we need at least one extra throw; and even then, half the time we must begin again. So $x = 2 + 1 + (x/2)$, hence $x = 6$.

In Penney-ante, if your opponent selects HHH, you should choose THH, and your winning chance is 7/8; if she picks HHT, again select THH, and win 3/4 of the time; to her HTH, respond HHT, against THH, use TTH – in both these cases, expect to win 2/3 of the time. Symmetry leads to the best choices against TTT, TTH, THT, and HTT.

References and further reading

Chapter 1

I. J. Good, *Probability and the Weighing of Evidence* (Griffin, 1950). Jack Good worked with Alan Turing on the Enigma project at Bletchley Park, and later at Manchester University.

L. J. Savage, *The Foundations of Statistics* (Wiley, 1954). Jimmie Savage was a leader of the subjective approach to probability.

Chapter 3

W. Feller, *An Introduction to Probability Theory and Its Applications*, Vol. I (Editions, 1950, 1957, 1968). Unsurpassed in its influence.

I. Hacking, *The Emergence of Probability* (Cambridge University Press, 1975). Authoritative, highly respected.

A. N. Kolmogorov, *Foundations of the Theory of Probability*, 2nd edn. (Chelsea, 1956). (The original publication in 1933 was *Grundbegriffe der Wahrscheinlichtkeitsrechnung*.)

S. M. Stigler, *The History of Statistics* (Harvard University Press, 1986). Scholarly work, meticulously researched.

I. Todhunter, *A History of the Mathematical Theory of Probability from the Time of Pascal to that of Laplace* (Chelsea, 1949). The chapter on Laplace contributes almost a quarter of this comprehensive work.

Chapter 4

D. J. Hand, *Statistics: A Very Short Introduction* (Oxford University Press, 2008). Very readable account from a President of the Royal Statistical Society.

Chapter 5

B. Goldacre, *Bad Science* (Harper Perennial, 2008). Mainly about misuses of statistics in medical science, full of good sense.

Chapter 6

N. Henze and H. Riedwyl, *How To Win More* (A. K. Peters, 1998). The subtitle 'Strategies for increasing a lottery win' tells it all.

E. O. Thorp, *Beat the Dealer* (Vintage Books, 1966). Made Thorp famous, but his fortune came from the stock market, not casinos.

Chapter 7

J. Fan and R. A. Levine, 'To Amnio or Not To Amnio: That Is the Decision for Bayes', *Chance*, 20(3) (2007). The full account, outlined in this book.

RAND Corporation, *One Million Random Digits, with 100,000 Normal Deviates* (RAND, 1955). Contains exactly what it says on the tin.

Chapter 8

Significance, 2(1) (2005). This journal contains several useful articles about probability and the law, including Peter Donnelly's 'Appealing Statistics', and Tony Gardner-Medwin's 'What probability should a jury address?'.

Websites and other publications

Several public-spirited groups and individuals post or update material germane to probability on the internet. In no particular order, I recommend:

http://www.dartmouth.edu/~chance/ takes you to the Chance website of Dartmouth College, with a variety of useful links, including the archive of 'Chance News', video and audio material.

http://www.mathcs.carleton.edu/probweb/ is aimed at researchers and teachers, but others will also be pleased to see the list of books, newsgroups, pertinent quotes, and miscellaneous information.

http://www.plus.maths.org/ The very readable online *Plus* magazine frequently contains probability material among its articles.

For accurate information on gambling odds in a host of games, Michael Shackleford has set up http://www.wizardofodds.com/ where your query may well be answered.

Significance is the magazine of the Royal Statistical Society. No formal qualifications are needed to join the RSS, merely an interest in statistics.

Statistical Science is a mainstream academic journal; several issues contain highly illuminating 'conversations' with distinguished probabilists who talk candidly about their careers.

Index

A

B

C